The Forgotten Exodus: The Into Africa Theory of Human Evolution

Bruce R. Fenton

Dedication

My deepest gratitude to all my ancestors, especially my father, Roger, and my mother, Mary. My love and appreciation to my wife, Daniella, and our three children, Zackary, Ariane and Camilla.

Preface

This book started out as some background fact checking to support my suspicions of an Australasian connection to the builders of a mysterious megalithic site in Ecuador. Having uncovered a link between the *Lagoa Santa type* people of ancient Brazil, I followed the evidence deeper into prehistory, leading me to proof of a forgotten history of Australasian colonisation of America. This account conflicted with the familiar narrative of human origins that involved a settlement of our planet by men and women walking out of Africa 70,000 years before the present day.

My investigations of the mysterious archaeological site in the Amazon jungle involved direct expeditions, meetings with government officials and collaboration with international teams of experts. Part of this story featured in the UK's *Telegraph* newspaper when in a December 2013 they published an article titled *Explorers hot on the trail of Atahualpa and the Treasure of the Llanganates*. The growing international coverage of my activities culminated in me being appearing on the popular *Science Channel* show The Unexplained Files; the episode was on lost races of giants, but my video footage from the Ecuador expedition also aired. To say that this research has been an exciting journey would be a massive understatement.

Incredibly, my small background task of fact checking thea human origins and early migrations spiralled into a mind-blowing recognition that there was an entirely lost chapter in human evolutionary history. Discovery after discovery pointed me towards the stunning conclusion that the popular Out of Africa theory for human origins was wrong, and that the understood colonisation of our planet was utterly flawed. This book offers the results of my detailed investigation into early hominin history; it is perhaps the most important document I will ever write.

Bruce Fenton

Sydney, Australia, April 2017

Contents

Acknowledgments

This undertaking required input from a multitude of academics including paleoanthropologists, archaeologists, anthropologists, climatologists, anatomists, historians, biologists and many laboratory technicians. Though these persons are too many in number to be listed here by name, I offer my sincere appreciation to the community of researchers that made this book possible.

It was my greatest fortune that I had previously collaborated with the two Australian researchers, Steven Strong and Evan Strong. That joint project introduced me to their work on human origins and the possibility that the Out of Africa Theory was a flawed evolutionary model. Steven and Evan provided me with important avenues of research, forwarded several articles, and made me aware of the essential link between my investigation of ancient peoples in South America and humans of a similar period in Australasia. These two men should be acknowledged as the founders of an Out of Australia Theory for human origination. I look forward to seeing whether our respective findings in this research arena agree.

As ever my deepest gratitude to my beloved partner Daniella, for her forbearance, patience and support throughout this time-consuming project. There are, as ever, many other things I could be doing to help more at home instead!

My most sincere appreciation is extended to my friend David Thomas for his immense efforts in the editing of this manuscript. There is only so much that automated spelling checks and professional grammar applications can do, a pair of human eyes, in the head of a professional editor, remain an essential tool for writers. One day I will find a way to return the favour.

Thank you to my beta readers, Steven Hope and Tim Hawkins, I am very glad that you found my work interesting and worthwhile. To all my loyal readers and fellow researchers that have supported my work, a big thank you.

Foreword by Graham Hancock

The Forgotten Exodus is a timely, thought-provoking and extremely useful little book by Bruce Fenton. And while it is not the only, or even the first, book to propose an "out-of-Australia" evolutionary origin for anatomically modern humans, it is by far the best. Indeed, it has the potential – although I cannot promise that this potential will be fulfilled – to rewrite history.

Fenton's arguments, and the impressive body of evidence he has assembled, fly in the face of the very powerful – and for the most part uncontested and unexamined – scientific consensus that *Homo sapiens* evolved exclusively within the African continent and spread outwards from there within the last 70,000 years or so. Unquestioning acceptance of this "out-of-Africa" reference frame is to be found everywhere in the relevant scientific literature. To give just one example here, consider these remarks from the Australian Research Centre for Human Evolution which recently pronounced, as though retailing an established fact, that:

> "Over 50,000 years, ago, Australia was the final destination of the first great migration of modern humans out of Africa. Prior to reaching our shores, people entered the archipelago world of Southeast Asia, a vast and ecologically diverse region once host to humans of a far more ancient character: *Homo erectus* ('Java Man') and *Homo floresiensis* (the 'Hobbit'). When and how modern humans colo-

nised this region, the nature of their interactions with earlier hominin lineages, and the pattern and timing of the demise and extinction of the latter, are fiercely debated."

It's right and proper that there should be "fierce debates" around such important subjects. The one thing that is not "fiercely debated", however – indeed it is barely discussed at all these days – is the ruling "out-of-Africa" paradigm which has been in place, and uncontested, for so long that most scientists seem to have forgotten it is still only a theory and not a body of established and irrefutable facts.

Yet if Australia was indeed the "final destination" of humanity's primordial trek out of Africa, isn't it odd that the Australian continent offers us some of the *oldest* evidence in the world of an anatomically modern human presence, as well as evidence of other human and pre-human species? Moreover, as Fenton demonstrates, this context of great human antiquity *very far* from Africa is not limited to Australia but extends across much of East and Southeast Asia as well. The arrival of *Homo sapiens* in Europe appears to have been relatively recent by comparison although intuitively, if "out-of-Africa" is correct, one would have expected African migrants to colonise nearby Europe long before they reached far-off Australia. Likewise, recent research has demonstrated conclusively that the modern human genome contains stretches of DNA contributed during episodes of interbreeding with our close evolutionary cousins the Neanderthals and the Denisovans – and again the strongest signals of this are not found in Europe – close to Africa – and certainly not in Africa itself, but far away in Australasia.

Bruce Fenton maintains that the fog of confusing and counterintuitive data that presently surrounds the story of human origins will not clear until modern science overcomes the limitations of the out-of-Africa reference frame and allows itself to consider the possibility that Australia was not the "final destination" of human migrations but their source.

That is bound to seem to many like a hypothesis so outrageous that it should not even be considered.

But before making up, or closing, your mind I urge you to read this intriguing book, look closely at the case Fenton makes, and double-check his references. In the process you will discover that many supposedly firm and established "facts" about human origins are by no means as well founded in the solid bedrock of supporting evidence as they should be, given the weight of interpretation put on them. And you will be confronted by exciting new evidence, never properly considered by the mainstream, that suggests we may be poised on the edge of a paradigm shift that will affect many different disciplines for years to come and require a profound rethink of who we are and where we came from.

Graham Hancock

Bath, England, April 2017

Hominin Timeline

Approximately fifty-five million years ago, the very first *primates* appeared on Earth, these are ancestors of humans and apes. Around six million years ago, a female primate gave birth to the first of the hominin line, this mother is known as the Last Common Ancestor (LCA). The obvious question here is, when did these primates become people? We know that chimps are intelligent, but we also know that they lack our rocket-to-the-moon level of thinking. There had to be a moment in time when the LCA gave birth to children that were solely hominin ancestors. The current dating for this divergence event is around 7 million-years-ago (Mya).

We need not examine the complete details of every species of hominin and pre-hominin form from seven million years; that would be an enormous project. We will take a quick look at some of the most important hominins living in Africa between the theorised divergence event for primates and hominins and the arrival of larger brained hominids around 2 Mya. In later chapters, we will separately examine several more recent hominin forms living after 2 Mya.

This is believed to have been one of the first forms to emerge after the divergence from our primate ancestors.

Name: *Sahelanthropus tchadensis*

Period: 6.8 – 7.2 Mya

Location: Chad

Morphology & Characteristics: Centrally placed foramen magnum, suggestive of *bipedal* movement. Small, human-like, canines, suited to an omnivorous diet.

Brain size: 350cc

Fossils: Skull fragments & teeth

Name: *Orrorin tugenensis*

Period: 6 Mya

Location: Kenya

Morphology & Characteristics: Comparative studies of thighbone shaft strength and size of the joints found a marked similarity to hominins living post 4 Mya, suggesting bipedalism.

Brain size: 350cc

Fossils: leg bones

Name: *Ardipithecus ramidus*

Period: 4.4 Mya

Location: Ethiopia

Morphology & Characteristics: A relatively short, broad, centrally positioned cranial base and related modifications of the intersection between skull and spine. The cranial base of modern human beings differs profoundly from that of all other primates, and therefore common features in that part of the anatomy suggest an ancestral relationship.

Brain size: 350cc

Fossils: Skull

Archaeological sites across Eastern Africa have revealed fossils from what appear to be several distinct species of a genus known as Australopithecines, at least seven are officially recognised; *A. afarensis*, *A. africanus*, *A. anamensis*, *A. bahrelghazali*, *A. deyiremeda*, *A. garhi* and *A sediba*. Offered here are the details of only the earliest and the most recent forms of this genus.

Name: *Australopithecus afarensis*

Period: 4 - 3 Mya

Location: Kenya and Ethiopia

Morphology & Characteristics: Exhibits the anatomy of an able biped but with retained tree climbing abilities and the potential for quadrupedal locomotion. Evidence of evolution towards exclusively bipedal movement is observed in adaptations to the toe bones.

Brain size: 420 - 500cc

Fossils: Skeletal fragments

Name: *Australopithecus sediba*

Period: 2 Mya

Location: South Africa

Morphology & Characteristics: Transitional features between archaic and modern hominins, most notable are the highly dexterous hands, well suited for utilising stone tools.

Brain size: 450cc

Fossils: Complete skeletons

Name: *Homo habilis*

Period: 2.4 – 1.5 Mya

Location: Kenya, Tanzania

Morphology & Characteristics: The greatest distinguishing features are several changes in the jaw, teeth and hand shape that set it apart from earlier hominins. *Homo habilis* (handy man) was the first ancient hominin to be strongly linked to stone tools.

Brain size: 610 – 800cc

Fossils: Skeletal fragments

Name: *Homo naledi*

Period: Undated (Somewhere between 2.5 – 1 Mya)

Location: South Africa

Morphology & Characteristics: Transitional features between archaic and modern.

Brain size: 560cc

Fossils: Complete skeletons

Name: *Homo rudolfensis*

Period: 2 Mya

Location: Kenya

Morphology & Characteristics: Modern Human-like facial features combined with archaic teeth. Suggested as possibly a large-brained Australopithecine rather than a *Homo*.

Brain size: 750cc

Fossils: Section of skull and brain case

Name: *Homo erectus*

Period: 1.9 Mya – 35 Kya

Location: Africa, Asia, Europe

Morphology & Characteristics: Over the period of existence this hominin evolved from a robust and relatively small-brained being to a more anatomically modern form with a large brain. Many features of modern humans emerged first among *H. erectus* and it may be responsible for the first art, first language and first sailing.

Brain size: 750 - 1000cc

Fossils: Complete skeletons

Introduction

The title of this book should give a good indication of the core theme; it offers a bold new theory of human origins that includes positioning the first emergence of *Homo sapiens* in a location outside of Africa, with a later exodus that takes modern humans into that continent. This book is much more than a revised story of *Homo sapiens* evolution; it is also a completely updated analysis of hominin evolution from seven million years ago, up until around twelve thousand years before the present. Included in this work is a discussion of other modern hominin forms, such as Neanderthals and Denisovans. It is intended to be accessible to anybody with a keen interest in human origins, evolutionary sciences, and the ancient migration routes of the very first fully modern humans. This book is not about the historical period of human activity; there is no discussion of civilisations and little reference to human cultures.

The reader will learn that many assumptions made in the area of human origins lack the strongly supportive evidence we should expect, while other commonly-claimed conclusions are fundamentally flawed. During the last few decades scientists have developed several new methods for dating fossils directly, and for dating material associated with ancient hominin habitation sites: some of these are uranium-lead dating, thermoluminescence dating, potassium-argon dating, and an improved accuracy for radiocarbon dating. Scientific researchers today also have the benefit of powerful supercomputers, allowing us to visually roll back past climates and even observe usually slow geological changes in a short video presentation. It is these exciting developments at the cutting edge of palaeoanthropology that have allowed me to collate strong evidence for a new and improved theory of our human origins.

The intention of this book is to update, and where necessary correct, the existing scientific consensus model of human evolution and early hominin migrations. You will find that in some areas I am merely making mild revisions, in other parts we will be thoroughly rewriting vast chunks of the current consensus academic theory. Everything shared with you is factual to the best of my knowledge, being based on the work of the most esteemed scientists in the fields of evolutionary biology and palaeoanthropology. Wherever possible, we focus on peer-reviewed scientific papers based on research guided by solid protocols, in almost all cases the scientific sources I have used can be considered mainstream authorities. There are a few instances where non-academics are my sources, but only where they offer unique insight into research that is itself accepted by the scientific community. The revised view of human origins on offer here is so radical that I could not risk undermining it by using potentially questionable sources, even where such sources appeared to provide a genuine revelation which would support this work.

Revision of our understood human origins starts at the very beginning of **Chapter One**, exploring the divergence of the human line from that of chimpanzees, around 7 million years before the present, calling into question even the basic anatomical form of the infamous missing link. During the evaluation of data in the first chapter, we revise our family tree, cutting off a sizeable number of branches in the process. The first members of our genus, *Homo*, are shown to have lived millions of years before the consensus narrative had previously led the public to believe. The chapter ends with two new hominin names and a very different start to the human story.

As we move onwards through the evolutionary journey, **Chapter Two** gives us an encounter with new *Homo* forms that start to closely resemble modern humans, *Homo heidelbergensis*, *Homo neanderthalensis* and *Homo antecessor*. We ask some brave new questions of these hominins. Could any of them be our real ancestors? Are the current hominin classifications all legitimate species? These questions are well worth asking, and the answers may well shock some readers. Before this chapter is over, we find ourselves facing a much earlier emergence date for archaic *Homo sapiens* and a severe shortage of viable ancestors.

The investigation takes a turn for the bizarre as we enter the Asian region in **Chapter Three**. We discover a new human relative, the Denisovan, while our old friend the Neanderthal reveals a mysterious family secret about our shared homeland. The real story here is a series of mind-blowing archaeological discoveries made in China over the last few years: *Homo sapiens* fossils older than any in the Levant and almost as old as those of African *Homo sapiens idaltu*. Could the Chinese academics be justified in their claim that modern humans evolved first in China?

We leave China now on the trail of the Denisovans, but to find them we have to take two important journeys in **Chapter Four**. Our first excursion is to the far past, to the time of *Homo erectus* while the other journey is a fast-flowing stretch of deep ocean. To track down the Denisovans, we are forced to rethink the first emergence of two key human traits: communication and advanced engineering. Was *Homo* *erectus* more like us than we ever dared to imagine? A single engraved shell calls everything we thought we knew suddenly into question.

Chapter Five begins as we penetrate the impenetrable barrier, taking us into the domain of the Denisovans, only to discover they are far from alone. How many hominins inhabited prehistoric Australasia? It seems that no academic can answer this question with any certainty, beyond admitting that it was a higher number than anybody believed. Strange things are afoot down under; on the continent typically thought of as almost the last landmass visited by any hominin species we find plenty of evidence contrary to this position.

In **Chapter Six**, we find ourselves pushing the last pin into the map; the first emergence of modern human beings right here in the Australasian continent. Can it be that we have found the real site-zero for the human story? The scientists confirming our findings are of the very highest pedigree, and the archaeological sites uncovered only strengthen their claims. It is an extraordinary claim, and extraordinary claims require extraordinary evidence.

There is nothing quite as taxing for me as attempting to understand molecular biology, but if we are to accept ancient Australasian Aboriginals as the founder population for all humans on Earth, we need to see the genetic data. The geographic spread of Y-chromosomes and mtDNA lineages are the matter at hand in **Chapter Seven**. Is the resulting evidence of global catastrophe and an incredible exodus the extraordinary final proof we required?

All journeys must come to an end. In our case, that end comes as we move back into Africa, twice in fact. From where our investigation began, so it comes to a close. **Chapter Eight** reveals the long-lost story of a Forgotten Exodus and the human migration into Africa. Loose ends are tied up as we reconsider *Homo sapiens idaltu* and the mysterious origins of the sub-Saharan KhoiSan peoples in **Chapter Nine**. This is only the end of the first stage of our journey into a lost human history, the Forgotten Exodus has more to reveal.

Chapter 1 – The emergence of *Homo Australopithecus*

Seven million years ago in Africa, a population of mysterious creatures gave birth to offspring that would eventually evolve into both chimpanzees and modern human beings (*Homo sapiens sapiens*). Evolutionary scientists claim that these ancestral beings were quadrupedal arboreal primates. It will perhaps surprise you to know that the existing evidence equally supports the possibility that these ancestors were fully bipedal ground dwellers. A bipedal ancestor as the missing link may sound like a bold claim, but it has a highly-respected academic at its source, Professor Alice Roberts (anatomist, osteoarchaeologist, physical anthropologist and palaeopathologist). Professor Roberts gave a presentation in 2012, *Origins of Us: Human Anatomy and Evolution*, in which she made a remarkable comment, "maybe the last common ancestor looked like us, and it's chimpanzees that have changed."

Professor Roberts has questioned several assumptions made about the missing link between humans and primates; she notes that there is no particular reason why a quadruped has to become bipedal if it renounces life in the trees, nor need it give up a balance-enhancing tail. It is also clear to her that modern primates move in a most peculiar fashion, quite unlike other quadrupedal animals: typically primates run on their knuckles. On the matter of the last common ancestor, it begins to appear that a fundamental reassessment of the details may be required.

We are not yet able to fully test the theory that *Homo sapiens* had a human-like ancestor living seven million years ago, rather than an ape-like ancestor. I will openly admit that this is a suspicion I share. Professor Roberts also points out that the first three possible ancestors, *Sahelanthropus tchadensis*, *Orrorin tugenensis* and *Ardipithecus ramidus*, might very well be ancestors of modern chimpanzees rather than of human beings. I suspect that Roberts is correct and that we are yet to find any fossils of human ancestors living before 4.4 Mya. Our earliest known ancestors are the *Australopithecus* hominins. As I will now show you, fundamentally human characteristics extend deep into the formative era of the *Australopithecus*, but these same human traits are not associated with fossilised beings from any earlier age.

On our list of ancestors provided in the Hominin Timeline, we note that the first to be accepted as a member of the genus *Homo*, 2.4 Mya, is *Homo habilis*. This early ancestor gained its place on the family tree by having an association with tool use. The creation and use of stone tools are considered the core standard for acceptance into our genus; the same level of consideration must apply to all known hominins. To prove my claims about *Australopithecus* being members of our *Homo* genus I need to offer substantial evidence of tool use amongst these more archaic hominins. Incredibly, this proof of previous tool use already exists.

Recent analysis carried out on a fossilised horse femur, from a site named Bouri, Ethiopia, revealed evidence of damage consistent with hominins using stone tools to dismember and fillet the horse's leg. This bone was discovered in proximity to further fossilised bones bearing similar marks, offering conclusive evidence of hominins using stone tools 2.5 Mya. At the time of this horse being dispatched the only hominin form proven to be present in that region was *Australopithecus garhi*.

Further support for lengthening the timeline of *Homo* came from the Ledi-Geraru archaeological site, located in Ethiopia's Rift Valley region. It was at Ledi-Geraru, in 2013, that a partial jawbone from a hominin displaying modern morphology pushed the origin of the *Homo* genus back to at least 2.8 Mya. This dating is positioned well before the time of the *A. sediba* individuals uncovered at Malapa (see Hominin Timeline). The Ledi-Geraru fossils also displayed several transitional features, having both the receding chin line of *A. africanus* (which vanished from the fossil record around 3 Mya) and the slim molars and mandible shape more typical of recent *Homo* forms. It is possible that the bones may be related to interbreeding between *A. africanus* and the earliest forms of *Homo*. Fred Spoor, professor of evolutionary anatomy at University College London, noted that the Ledi-Geraru jaw might yet offer "a plausible evolutionary link between *Australopithecus afarensis* and *Homo habilis*".

The discoveries that *A. garhi* were using tools and that *A. afarensis* might be closely related to *H. habilis* are major revelations, but they do not take us back far enough to be conclusive evidence for the theory I have suggested. It is only at the Lake Turkana archaeological site in North West Kenya that we find the earliest signs of stone tools being used to cut meat. The Lake Turkana discoveries produced dates far more radical than those associated with Bouri. Tools used at this second site are 3.3 million years old. Speaking on the Lake Turkana discovery, Richard Potts, the director of the human origins program at the Smithsonian's National Museum of Natural History, voiced the revealing opinion that the "capabilities of our ancestors and the environmental forces leading to early stone technology are a great scientific mystery."

These recent discoveries, when considered together, suggest that we should no longer be thinking of various species of *Australopithecus* living before the emergence of the genus *Homo*. It is my view that from 4.4 Mya, we see the arrival of a new hominin, one that I would name *archaic Homo australopithecus*. What we currently know as *H. habilis* is in fact as a transitional form of *Homo australopithecus*. *Homo australopithecus* eventually gave way to the more modern form we know as *Homo erectus* (upright man). This transition between forms would have taken considerable time; as best as we can speculate, this process was complete by two million years ago. The variation observed in early hominin fossils supports the existence of what we today call racial types, individuals within a species that have developed adapted morphology rather than examples of distinct species. We will soon return to the discussion of a single evolving human lineage displaying high morphological variance, but with additional evidence from a slightly more recent era.

The commonly accepted geographic location for the first emergence of *H. erectus* is East Africa; the dating in official circles is 1.9 Mya. In my personal research, it has proved difficult for me to track down particular fossils in Africa accepted as *H. erectus* dated any older than 1.8 Mya. It may be that the details of the older fossils are simply not well publicised.

Homo erectus was fully bipedal. Many of his ancestors appear just as well adapted to climbing trees as to walking on the ground. Walking entirely on two legs is an anatomical adaptation that brings both benefits and hindrances. We know that for hominins the benefits gained from using upper limbs for tasks other than movement far outweighed any negatives (such as reduced speed or mobility). It is only with the evolution of skilful hands, able to manufacture tools, that we can recognise the first members of our genus.

The skull morphology of *H. erectus* displays some stark differences to that of *Homo sapiens*, including a significantly protruding brow ridge, a prominent jaw with huge teeth, and the absence of a chin. The most significant difference was in the average size of the brain case, with *H. erectus* having 750cc as opposed to *H. sapiens* with 1250cc. Despite the many differences, *H. erectus* was unarguably a human, having far more in common with us than do apes.

There is a theory espoused by some academics, including Harvard anthropologist, Richard Wrangham, that *H. erectus* was a master of controlled fire from as early as 1.9 Mya.

Wrangham suspects that the energy boost provided by cooking high calorie-containing meat played a significant role in hominins evolving stronger bodies and larger brain sizes. Larger brains and more powerful bodies both require sufficient fuel; roasting meat considerably increases calorie absorption during digestion. Controlled fire was considered to be a technology beyond ancient hominins until evidence from Wonderwerk Cave in South Africa proved that fire had been used for cooking purposes at least one million years ago, during the era of late *Homo erectus.*

Although we may be speculating about controlled fires, we do know that the stone technology of *H. erectus* was progressing forwards along with brain capacity. With increasing brain size comes increased problem-solving skills and new flexibility in thinking. This increasing cognition leads directly to a new strategy, expansion into unfamiliar territories beyond the African continent. Despite being thought of as an African hominin, *H. erectus* fossils are found more often at sites in Asia where they had settled at least as early as 1.6 Mya. In 1995 two reports emerged on a collection of fossils and stone tools discovered at Longgupo Cave in south central Asia. They offered a dating of 1.9 million years for the artefacts, based on magnetic and electron spin resonance. At Mojokerto, Indonesia, one *H. erectus* skull was dated to 1.8 Mya through radiometric dating methods.

Homo erectus groups in Asia display some morphological divergence away from those in Africa which initially led to the profound misunderstanding that they represented two separate species. Even today some scientists still refer to

African *H. erectus* as *Homo ergaster*. We now know that Asian and African *Homo erectus* were simply two races within a single evolving species. Each regional lineage on Earth today displays noticeable adaptations to the physical form which better suit local environments. We can readily appreciate that divergence into distinct local populations takes much less time than does separation into different species. We will return to the crucial matter of variance within the *H. erectus* species very shortly.

We can assume that the ancestors of Asian *H. erectus* passed through the Levant and then headed East towards Central Asia before turning towards South East Asia. It is in these regions that we find some of the oldest relevant fossil sites. There was, however, at least one other direction taken by these migrants, a more northerly route out of Africa. Between 1999 and 2001 *Homo erectus* fossils were discovered at Dmanisi, high up in the Georgian section of the Caucasus Mountain range. These Georgian fossils are at least 1.8 million years old, very close in time to the age of the earliest *Homo erectus* fossils discovered in Africa. The very fact that Georgian *H. erectus* were almost contemporary with the African fossils came as a major shock to the scientific community.

The international research team in Georgia led by David Lordkipanidze of the Georgian National Museum in Tbilisi included top scientists from Switzerland, Israel and the United States. The results of the analysis carried out on the skulls, among which was the first completely preserved adult hominin skull from the early Pleistocene, left the entire field of palaeoanthropology in disarray. The remarkable degree of morphological variation displayed across the five Dmanisi skulls immediately called into question the validity of numerous designated hominin species from between late *Australopithecus* to early *Homo*.

The paper published by the Georgian research team in the journal *Science* concluded that the skulls were strong evidence of a single evolving lineage of early *Homo*, with a correspondence of form across geographic regions as a key feature of the species. The morphological variance seen among the skulls correctly matched that observed among African hominin fossils spanning as far back as 2.4 Mya, also of Asian and European fossils from between 1.8 to 1.2 Mya. The incredible findings of the Georgian research invalidated as many as nine designated species of hominin. The physical evidence from Dmanisi points towards a new understanding in which all hominin species from *A. Africanus* onwards are *Homo erectus*.

Lordkipanidze and his team had revealed that *H. erectus* extended much further back into the past, to at least 2.4 Mya. We have also seen that behaviour associated with members of our genus is now also associated with hominins currently classified as early *Australopithecus*. The Bouri and Lake Turkana archaeological sites offer further physical evidence supportive of the conclusions provided by the Dmanisi research team, that we are in fact dealing with a single global species of hominin, all members of which are *Homo*.

We now have a completely revised review of history in which modern humans evolved from a single evolutionary lineage, starting with an unknown example of archaic *Homo australopithecus*. It is probable that tools were a feature of our history from the very beginning (even chimpanzees use primitive stone tools) but were indeed being used by 3.3 Mya. This evolutionary line began a morphological transition from *Homo australopithecus* to *Homo erectus* around 2.4 Mya. *Homo erectus* would later go on to become a global species, with numerous racial groups displaying localised adaptations, but there would be no significant anatomical evolution of this species until after 1.2 Mya. The confusing of identities, such as Asian *H. erectus* and African *H. ergaster* as separate species, is an issue extending to various supposed hominin species.

With the early fossils of *H. erectus* in Asia and the Caucasus region, we find evidence suggestive of hominins having left Africa earlier than commonly believed. They may even have ranged in and out of Africa. To understand the next stage in our evolution we need to follow later generations descended from the *Homo erectus* population resident in the Caucasus region. It is in Europe that we find evidence of a subsequent stage in hominin evolution and bear witness to the demise of yet another hominin classification.

Chapter 2 - *Homo heidelbergensis* is a Neanderthal

Although we know that populations of *Homo erectus* were living on the Eurasian border around 1.8 Mya, as proven by the fossils discovered at the Dmanisi archaeological site, it still took a few hundred thousand years for this population to colonise Europe. The extreme climate in Europe acted as a barrier to early hominin migrants, severely limiting the number of habitation sites. The only archaeology available from 1.8 to 1.2 Mya is a scattering of stone tools found at sites in Russia, Bulgaria and France. The European fossil record is bare until 1.4 Mya. The first remnants from this time are a pair of fossilised molars uncovered at a dig in Orce, a site located in the province of Andalusia, Southern Spain. The teeth are those of a ten-year-old hominin, assumed to be a *Homo erectus*.

The oldest fossils from hominins in Europe are all found in Spain. After Orce, the second oldest site, dated to 1.2 Mya, is in the province of Atapuerca, this time in the far North of the country. The next site identified is located nearby in the same region, only dating to several hundred thousand years later. Before discussing these ancient sites in more detail, we need to understand why there are so very few hominin sites in Europe associated with this early period. The reason transpires to be something far more ominous than simply cold weather.

While studying ancient regions of the human genome a geneticist, Lynn Jorde at the University of Utah, stumbled upon evidence of a collapse in the numbers of people on Earth around 1.2 – 1.0 Mya. These revelations had come while the scientists were analysing mutations positioned near to ancient DNA segments known as *Alu markers*. These marker sites allow researchers to calculate levels of diversity among early humans. The cause of this near extinction event, known as a *population bottleneck*, had to be a previously unknown global catastrophe, one that had reduced the world's population of hominins (all known forms) to perhaps as few as 18,500 individuals. The research team published their results in the *Proceedings of the National Academy of Sciences* (2010).

A small hominin population required few habitation sites; the chances of us finding any of these, when spread across the European continent, would be very slight. There are also significant evolutionary considerations associated with small numbers of breeding pairs; one obvious consequence is lower genetic diversity. Small populations often preserve the existing levels of diversity until numbers recover. If any significant mutations occur while population numbers are small, these will be passed on to a huge percentage of individuals. Extensive diffusion of genetic mutations can be extremely problematic if they are detrimental as they might lead to outright population extinction. If a highly beneficial mutation becomes widely distributed, this could potentially accelerate the evolution of an entire species. We should keep this understanding in mind as we proceed with our research journey in Spain. This may well also be important in respect to understanding the divergence of archaic *H. sapiens* and other hominins from their shared ancestors living during the time of low population numbers.

The archaeological site in Northern Spain, given the name *Sima del Elefante* (pit of the elephant), produced a single piece of jawbone along with several teeth. On its outside surface, the jawbone shows features found in earlier fossils of *Homo erectus*, while on the internal surface it is lightly built, this being a feature of more modern humans. The remains of Spain's second oldest resident were calculated to be 1.2 Mya. With only fragments uncovered, a definitive attribution to a known hominin species is virtually impossible. The discoverers assumed the bones must be from an early member of a nearby population given the designation *Homo antecessor*. The attribution of the fossils at *Sima del Elefante* is not very convincing. There remains to this day lots of controversy surrounding the recognition of *Homo antecessor* as a separate hominin species.

Homo antecessor is a uniquely European species, first recognised at the Gran Dolina cave site less than 100km from *Sima del Elefante*. The six hominins uncovered at Gran Dolina are now known to date to around 800 Kya. The discovering team, Eudald Carbonell, Juan Luis Arsuaga and J. M. Bermúdez de Castro, claim that unique morphological features observed in the fossils, especially in the nasal region, are proof that the fossils represent a new hominin species. Since the discovery, several scientists have even suggested that *Homo antecessor* may be a shared ancestor of both Neanderthals and *Homo sapiens*. This claim needs to examined.

Controversy still surrounds the decision to acknowledge the Gran Dolina finds as representatives of a new human species. The associated fossils are mostly from infants; this is problematic as children change appearance significantly post-puberty. Many highly distinguished researchers suspect that, had these youngsters developed to adulthood, they would have looked very much like other early European hominins, perhaps *Homo erectus* or an early form of *Homo heidelbergensis*.

Homo heidelbergensis plays a vital role in the commonly accepted scientific model of human origins, *Out of Africa* theory (OoAT). The OoAT states that around 600 Kya *H. heidelbergensis* emerged from among the populations of *Homo erectus* living in Africa, Europe and Asia. African populations of *H. heidelbergensis* are given a unique identifier, *Homo rhodesiensis*, despite the fact that the two forms display only superficial morphological differences. It is best to think of African and European *H. heidelbergensis* as racial groupings within a single species.

The *Sima del Elefante* fossils certainly exhibited some interesting adaptations, just as had the jawbone from Gran Dolina, but it is debatable whether there is enough to claim a new species. Would researchers claim these fossils represent a new species if they had had already known of the Dmanisi skulls? The findings from the Georgian research best fit a scenario in which *H. antecessor* would be considered European *H. erectus*. The Georgian research team specifically stated that the morphological variance observed at Dmanisi accommodates all known European fossils dated between 1.8 – 1.2 Mya. This date range includes the jaw from *Sima del Elefante*. The later fossils from Gran Dolina make the most sense as evidence for a transitional form between *H. erectus* and *H. heidelbergensis*.

Despite the glaring weaknesses in the argument for *H. antecessor* as a valid species, a few scientists have speculated that it might be a good candidate for the last common ancestor to both *Homo sapiens* and *Homo Neanderthalensis*, our closest relatives in the fossil record. The commonly accepted theory posits that a group of *H. erectus* in Africa evolved into *H. antecessor* before migrating into Europe. More recently they evolved into *H. heidelbergensis*. The population of *H. antecessor* remaining in Africa later evolved into *H. Rhodesiensis* and eventually became *Homo sapiens*.

Unfortunately, at least for those who support the *H. antecessor* migration model, we now have the Dmanisi fossils as definitively contrary evidence. The first hominins that entered Europe were clearly still *Homo erectus*, members of the same species also found in Africa and Asia. There is also no evidence of *H. antecessor* living in Africa. If *H. antecessor* truly is a unique species of hominin, it is also uniquely European. In fact, the only place it lived was Northern Spain.

We do not need to travel far to continue the story of hominin evolution in stone-age Europe. Within 200km of the previous two archaeological sites, we find a third cave site:

Sima de los Huesos (Pit of Bones). The fossils at *Sima de los Huesos* included the remains of twenty-eight individuals, dated to 430 Kya. Based on the existing logic we would assume that these must be the later offspring of our *H. antecessor* population. They are just down the road from the two previous sites. The archaeologists noted that the remains appeared to be those of *Homo heidelbergensis*, a species considered to be transitional between *H. erectus* and both *H. neanderthalensis* and *Homo sapiens*. The real identity of the species remained at question until more intensive investigation completed.

Two very different scientific research projects would go on to examine the bones from the *Sima de los Huesos* site. One project, focussed on ancient genetics, involved a team of scientists from the Max Planck Institute for Evolutionary Anthropology, headed by a molecular biologist named Matthias Meyer. The second project, focussed on the comparative morphology, involved scientists associated with Indiana University, led by the evolutionary biologist Aida Gómez-Robles.

Meyer's team began with the extraction of mitochondrial DNA (mtDNA), this being the genetic material passed down exclusively through the female line. Several interesting finds emerged from this initial study, some which we will discuss later. The most exciting revelation was that the fossils at *Sima de los Huesos* appeared to represent yet another new species, neither *Homo heidelbergensis* nor *Homo antecessor*. The new species was tentatively assigned the name *Sima de los Huesos hominins*. We need to recognise that the extraction of mtDNA from fossils almost half a million years in age was an enormous step forwards in hominin genome research. Most of the scientific community believed that the removal of viable material from the cell nucleus – nuclear DNA – would be impossible. Incredibly, Meyer's team would later announce that his team had successfully managed to do the impossible: they had recovered viable nuclear DNA.

Analysis of the nuclear DNA convincingly proved that the *Sima* hominins were, in fact, early Neanderthals. The presence of Neanderthals at such an early point in the evolutionary timeline pushed back the divergence of *Homo sapiens* from their human cousins quite considerably; the team calculated that the split had to have happened no more recently than between 550 to 765 Kya. The earliest associated fossils for *H. heidelbergensis* enter the records only after 600 Kya. Most finds are considerably younger. The revised dating for the divergence event was now too far back for *Homo heidelbergensis* to remain a viable candidate as last common ancestor for both species; it was now clear that they were just early examples of Neanderthals.

The Indiana University study team had already completed its attempt to identify the last common ancestor of both *H. sapiens* and Neanderthals through studying the comparative morphology of ancient European hominin fossils. The scientists involved had performed detailed analyses of dental bones and fossil teeth – over 1200 molars and premolars – with many individual hominins represented as well as 13 hominin forms (this being before the Dmanisi findings).

The Indiana University study published its results in the journal *Proceedings of the National Academy of Sciences* in 2013.

The hope was that with such a widely comprehensive analysis it would be possible to identify the shared ancestor of Neanderthals and modern humans firmly. All European hominin species are understood to have a representative population in Africa. Gómez-Robles stated that "Morphometric analysis and phylogenetic statistics were used to scrutinise every detail of the dental morphology of the last "possible" ancestor of humans and Neanderthals."

Despite all their dedicated efforts, Gómez-Robles and her team concluded that none of the available samples belonged to a line of hominins directly preceding modern humans. The scientists involved also declared that all the dental samples from Europe, and therefore all known European hominins, were associated with the Neanderthal lineage. These conclusions went even further than the results of the DNA research carried out on the *Sima de los Huesos* fossils in that they also dismiss the *Homo antecessor* as a possible candidate for our shared ancestor. Based on the dental evidence, the Indiana University study placed the beginning of the divergence of Neanderthals and *Homo sapiens* at around one million years ago, earlier even than the dating in Meyer's results and well before the time of the hominins resident at *Sima del Elefante*.

Homo heidelbergensis in Europe and *H. rhodesiensis* in Africa are so similar that after the findings of the DNA and fossil comparison studies it was no longer rational to suggest that the African variant could be ancestral to modern humans.

With the demise of *H. heidelbergensis* and *H. rhodesiensis* as possible ancestors for modern humans, a glaring gap appears in our family tree: there is currently no other hominin candidate identified in either the African or European fossil records that could represent an ancestor for modern humans. The paleoanthropologist Maria Martinón-Torres of University College London discussed the difficulties presented by Mayer's genetic data in an additional article published by *Nature* in 2016. Speaking on the matter of a viable ancestor for *H. sapiens,* Martinón-Torres was quite sure that researchers "should now be looking for a population that lived around 700,000 to 900,000 years ago." Martinón-Torres suggests *Homo antecessor* as a candidate species but she clarifies that this would require the discovery of related fossils in Africa or the Middle East.

In some respects, the suggested dating for *H. sapiens'* divergence away from ancestors shared with other hominins fits pre-existing data as paleoanthropologists already know that *H. erectus* fossils display only cosmetic morphological change between 1.8 Mya until 900 Kya. The fossil record shows us that by around 800 Kya there is a sudden and rapid expansion in hominin brain size, as well as several new adaptations to the anatomy. Knowing that the global population of hominins was tiny at this crucial time in history, we can also better understand how evolution across the species could occur at such breakneck speeds as are shown in the fossil record. What is less clear is what might have caused them.

Our research in Europe has produced revelations of considerable magnitude; we now have no possible ancestor for *H. sapiens* among European hominins, or among known hominin populations in Africa. The hominin usually identified as *Homo heidelbergensis* is shown to be nothing more than an archaic Neanderthal, just as many academics had long suspected. The commonly accepted dating for the divergence of fully modern humans from Neanderthals at around 500 Kya is found to be definitively wrong and probably underestimated by at least 200,000 years. Paleoanthropologists find themselves tasked with identifying a suitable ancestral hominin population that was living around 900 Kya.

There is extensive archaeological evidence of Neanderthals inhabiting Western and Central Asia, and several fossil finds have been ascribed to *Homo heidelbergensis* (archaic Neanderthals) further to the East, in China. As the oldest fossils of early Neanderthals are all found outside of Africa, it makes sense to follow this trail Eastwards through Eurasia in the hope that we may locate the place of their genesis. If we can find the hominins that gave birth to archaic Neanderthals, then we will have also found the population that gave birth to archaic *Homo sapiens*. It is time to head east of the Caucasus as we follow the hominin trail into Asia.

Chapter 3 – Fully Modern *Homo sapiens* in China 180,000 years ago

The nearly universal consensus in palaeoanthropology includes the belief that groups of *Homo heidelbergensis* (archaic Neanderthals) migrated from Africa to Europe around 600 Kya. There have always been serious flaws in this migration theory. The oldest fossils categorised as *Homo heidelbergensis* come from Europe. Almost no traces of Neanderthal DNA are present in members of modern African populations. We must remember here that Neanderthal DNA exists at significant levels among all non-African modern humans. With the revelations of the previous chapter, we can now understand that any early Neanderthals in Africa must have migrated into the continent. Judging by the best evidence available today, we can deduce that the small number of Neanderthals in Africa became extinct before the emergence of *Homo sapiens*. The extinction of African Neanderthals would explain why there was no interaction between the two species in that region and why we do not find any more modern Neanderthal forms in Africa.

So where did these first Neanderthals originate, if not Africa?

As Neanderthals are our closest relatives in the fossil record, sharing an ancestor with us until at least 900,000 years ago, anything we can learn about their evolutionary story will assist the investigation of our own. We have already investigated whether the shared ancestor might have been in Europe but have found only archaic Neanderthals and late era *Homo erectus*. Most paleoanthropologists consider Neanderthals a European species; this claim has turned out to be only partially correct. A paper on the genetic turnover in Neanderthals, published in *Oxford Journals – Molecular Biology & Evolution (2012)*, offered a revolutionary understanding. The research team behind the study led by Love Dalén noted that later Neanderthals lacked the diversity of preceding Neanderthals found in Western Europe. Strangely, for a supposedly European hominin, Central Asian Neanderthals displayed increased diversity. The data seemed to paint a picture in which waves of Neanderthals were migrating into Europe from somewhere in western or central Asia. This finding suggested that there was a larger and more genetically diverse Neanderthal population living somewhere in Asia, assumedly their ancestral population.

As we move our focus towards Central Asia, we find ourselves hearing of a truly incredible Neanderthal site. In 2010 a team of researchers announced the discovery of hominin remains inside a cave located in the Altai mountains of Siberian Russia. Russia is an enormous country; it is then perhaps helpful to mention that the Altai mountain range sits close to the Russian borders with Kazakhstan, Mongolia and China. The Altai cave site produced evidence of inhabitation by hominins ranging back to as early as 180 Kya, with Neanderthals being the original occupants. As the archaeological research progressed, it became apparent that modern humans had also been present at the site. Incredibly, it later transpired that some of the fossil remains were from a third human form previously unknown to science. This third species has since been tentatively named Denisovans (after a Russian hermit named Denis, who had formerly lived in the cave).

The remains of three Denisovan individuals were found in the cave; the youngest were those of a female child, dated to around 50 Kya. We even know, from extracted DNA, that the girl was tanned, with brown eyes and dark hair. The fossilised remains of the two older Denisovans in the cave date to at least 110 Kya. Although this is rather close to modern times compared to the period of our fundamental interests, detailed DNA analysis revealed that this human species had existed for hundreds of thousands of years. The results of the DNA study, carried out by the Max Planck Institute for Evolutionary Anthropology, determined that Denisovans were closer relatives of Neanderthals than were *Homo sapiens*. In respect to the divergence of these three modern human forms, Svante Pääbo, lead researcher for the study, observed from the results that "Denisovans began to diverge from modern humans regarding DNA sequences about 800,000 years-ago", which is surprisingly early in the timeline. The final shock was that the Denisovan genome also carried genes from a fourth, even older, hominin lineage that is undetected in the fossil record.

The close ties between Neanderthals and Denisovans would later be reconfirmed amongst the results of the *Sima de los Huesos* genetic research project. Initial testing of mtDNA extracted from the fossils at *Sima de los Huesos* had revealed an unexpected link between the Neanderthals and Denisovans. The final round of tests involving nuclear DNA had confirmed the validity of that finding. Upon confirmation of the unexpected relationship, Meyer had commented that the ancestors of the two groups carried mitochondrial DNA that is reflected in both, though not present in later Neanderthals. Meyer's findings suggested that the link between Neanderthals and Denisovans was both ancient and non-European – further confirmation that the ancestral populations of archaic Neanderthals had migrated into Europe – rather than having evolved there directly from *Homo erectus*.

Based on the combined archaeological and genetic data, it appears that the ancestors of Neanderthals and Denisovans had begun their split from the ancestor shared with *H. sapiens* sometime between 1 Mya and 800 Kya, with Denisovans and archaic Neanderthals diverging sometime later. The data might also support a scenario in which Denisovans separated away from the shared lineage first, with Neanderthals following soon after, and then *H. sapiens*. We must understand here that once our ancestors had parted ways with these other hominins, 800 Kya, the *Homo sapiens* lineage officially came into existence as a separate population destined to become fully anatomically modern humans: *Homo sapiens sapiens*.

Though the ancestors of all European hominins carried Denisovan genes, later generations no longer display this connection, suggesting that archaic Neanderthals must have been living within proximity of the archaic Denisovans sometime before the migration into Europe. The location of the ancient cohabitation must be somewhere in a more eastern territory. Even though inhabitants of the Altai cave included *Homo sapiens*, Neanderthals and Denisovans, the finds are all too young to represent evidence of our ancestral population.

If we were to locate the home territory of the Denisovans rather than that of the Neanderthals, we might still solve our puzzle. Examination of the genetic profiles of modern humans inhabiting the region surrounding the Altai mountains further suggests that we are not in the right location. No modern human groups from anywhere in Central Asia, or even in Europe, carry Denisovan genes today, though they do all have Neanderthal genes. It is once we evaluate the East Asians that we find traces of Denisovan genetic material, alongside a very significant contribution of Neanderthal genes. In the research paper *Higher levels of Neanderthal ancestry in East Asians than in Europeans* (2013) results of a study carried out by the Institute for Human Genetics led by Jeffrey Wall concluded that East Asian populations typically carry as much 40% more Neanderthal genes on average than modern Europeans. The results of this investigation were noted to be consistent with the findings of an earlier study associated with the Max Plank Institute, *Meyer et al. (2012)*, which involved a smaller population sample.

Among Eurasians, the Lahu and the Naxi, two Sino-Tibetan populations from China, are the closest living relatives to Neanderthals. North American Indians are the only people that carry more Neanderthal genes than the Lahu and Naxi. There is a mystery involved in the story of the North American Indians and their close relationship to Neanderthals, but it would take us off track were we to investigate it here and now.

The higher levels of genetic admixture among East Asians are the result of long and persistent interbreeding in that geographical region, on a greater scale than in Europe. The levels of archaic DNA among East Asians raises another question: to what extent did interbreeding occur between modern humans and Neanderthals in Europe? The prevailing consensus view incorporated into OoAT tells us that all interbreeding between the species happened as modern humans entered Europe. If there is evidence that the two populations intermingled largely in East Asia, this would be hugely supportive of the suspicion that archaic *Homo sapiens* shared a homeland with archaic Neanderthals somewhere in that region.

The first important piece of information that we receive on the relationship between Neanderthals and modern humans living in Europe is that it was either incredibly short or completely non-existent. While the ancestors of Neanderthals entered Europe hundreds of thousands of years ago, modern humans arrived very late, around 45 - 42 Kya. The research community broadly agrees that Neanderthals went extinct in Europe by around 40 Kya. Breakthroughs in the science of genomics, as well as new archaeological finds, collectively call for a rethinking of the old claims of significant interactions between Neanderthals and modern humans in Europe. The most intensive examination of the Neanderthal extinction event ever carried out revealed the impossibility of lengthy contact between them and modern humans. This forensic investigation involved the careful analysis of bone, charcoal and shell materials from forty archaeological sites spread across the breadth of Europe. After the analyses had been completed, the conclusion was that Neanderthals had vanished from Europe by between 42 – 39 Kya. Tom Higham, a radiocarbon scientist at the University of Oxford, said of these results, "I think that, for the first time, we have a reliable extinction date for Neanderthals."

With solid archaeological data now available dating the extinction of Neanderthals, and the subsequent arrival of early modern humans, it seems unlikely that they had sufficient time for notable levels of interbreeding in Europe. The remains of an early European modern human, Oase 1, uncovered in Romania, tells us that at least one group of modern humans interbred with Neanderthals close to the Eurasian border, around 42 – 37 Kya; this fossil evidenced around 6 - 9% Neanderthal ancestry. The problem with Oase 1 is that the individual in question was found to be no more European than Asian, and his lineage did not contribute any genes to modern European populations. Findings of the study of the Oase 1 remains were published in *Nature* (2015) and included a remarkable admission by the researchers that analysis of present-day genomes has so far failed to yield any evidence that Neanderthals interbred with modern humans in Europe. It appeared that Oase's people had gained their elevated levels of Neanderthal DNA while heading west through Asia, or perhaps right on the Eurasian borders.

Perhaps the most insurmountable flaw in the OoAT perspective on interbreeding in Europe is that the greatest amount of interbreeding between the two lineages, modern human and Neanderthal, is currently calculated to have occurred around 77 – 114 Kya. That date is incredibly early, especially when considering that *Homo sapiens* had not even left Africa by then, according to the consensus timeline for human origins. We already know that there is no evidence of Neanderthals in Africa at that time so we can't place the early interbreeding event on that continent instead. We know for a fact that Neanderthal women were bearing children for modern human men as early as 100 Kya, a fact that emerged from analysis of a neanderthal bone at the Denisova cave site, many thousands of kilometres from Africa.

Higham offers one possible lifeline for the OoAT, mentioning that "yet-unpublished data suggests the interbreeding events occurred about 55,000 to 60,000 years-ago", but this still places the interbreeding over 10,000 years before modern humans arrived in Europe. No matter how radical it may sound, interbreeding between *H. sapiens* and Neanderthals can only have occurred in Asia; there are no other remaining regions to consider. We do still need to pin down a part of the continent with both fossils of *Neanderthals* and *H. sapiens*. The only Asian region fitting the requirement is that of modern day China.

Two of the most famous fossil finds in China are the skulls known as the Dali skull and the Jinniushan skeleton, dated between 150 – 200 Kya, though exact dating is in question. Though the Dali skull has prominent brow ridges and some *H. erectus* traits, the consensus is that it is either archaic *H. sapiens* or *H. heidelbergensis* with a preference for the latter. The Dali skull closely resembles a transitional form between *H. erectus* and *H. sapiens*. The *Jinniushan* remains are even more distinctly modern, though apparently still exhibiting older features, they closely resemble the African fossils identifies as *Homo sapiens idaltu*, the brain case is slightly larger than that of most modern humans today at 1400cc. These fossils are most commonly considered to be evidence of an evolving population of local *H. heidelbergensis*.

A third, slightly more recent, Chinese fossil of equal relevance to us is the Maba skull from Guangdong province. The Maba skull dates to between 120 – 140 Kya. It is a partial skull displaying features in common with the skulls of Neanderthal and *H. sapiens*. The skull also exhibits unique features suggestive of a possible new and unknown hominin. The mysterious Maba skull seems to have commonality with further Chinese finds, especially teeth and bone fragments from Xujiayao, which also fail to match known hominin forms but have a high similarity to the fossils of the Denisovans. There is growing suspicion that at least some of the Chinese fossils represent either Denisovans or hybridised Denisovans. It may be that they are something entirely different, perhaps members of the mysterious hominin species detected in the genome of the Denisovans previously living in Siberia.

Fossils of ancestral Neanderthals from China are already well documented, but more recently compelling evidence has emerged for a presence of fully modern *H. sapiens*. The fossils attributed to modern humans in China date to the periods associated with both the earliest dated interbreeding event and the more recently dated interbreeding event. These fully modern humans in China are excellent candidates for the *H. sapiens* population that first carried DNA from Asian Neanderthals into Europe. Two archaeological sites, Lunadong Cave and Fuyan Cave, have both provided ancient fossils, chiefly teeth, which prove *Homo sapiens* were in these locations respectively at 70 - 127 Kya and 80 - 120 Kya. Two sites, both in Guizhou Province, have provided teeth identified as being from *Homo sapiens*; respectively they are three that are 112 - 178 thousand years old, and 47 that are 80 - 120 thousand years old. One paper reporting on teeth from Late Pleistocene Luna Cave concluded that the teeth from the archaeological site consolidate the proofs that firmly place modern humans in eastern Asia, between 50 - 120 Kya. Ironically, the same period now associated with modern humans in China exactly corresponds to a time long associated with a complete absence of people in East Asia.

Elsewhere a human jawbone recovered from the Zhiren Cave is especially fascinating not only because it is 100,000 years old but also because it includes a distinctly modern feature: a prominent chin. The only sceptical opinion on offer is that it might be that of a Neanderthal, but this would itself be interesting as it would undermine the claim that fully evolved Neanderthals were not present in China.

The dates for the first *H. sapiens* in China are problematic for the OoAT model. No *H. sapiens*, irrespective of whether they be archaic or entirely modern, should be settled in China earlier than the dates given for the human migration out of Africa. This inconsistency between the dating of Chinese hominin sites and the current consensus theory of human origins was the focus of discussion in a paper published by Wu Liu, Xiu-Jie Wu and María Martinón-Torres. Their statement on the subject was both direct and unequivocal: "Our study shows that fully modern morphologies were present in southern China 30,000–70,000 years earlier than in the Levant and Europe."

The oldest definitively *H. sapiens remains* from the Levant region are pieces of a 55,000-year-old skull found at Manot Cave in northern Israel. There is no way for the consensus model to account for more recent dates associated with humans living closest to the African exit and entry point; modern humans on the other side of the planet lived at a significantly earlier time. The Chinese finds are in some cases as much as 10,000 to 100,000 years older than the fossils of fully anatomically modern humans in Africa, and almost equal to the oldest of the archaic *Homo sapiens idaltu* fossils.

Some of the discoveries in China are considered suggestive of an ongoing process of evolution in a population of Asian *H. erectus* giving rise to *H. heidelbergensis* and perhaps other more modern forms of hominin. There are reasons to believe that this is at least part of the story, even if other crucial evolutionary events happened at some other geographical location. The evidence of *Homo erectus* evolving towards Neanderthals in East Asia offers further proof that we are moving towards our founder population.

Dennis O'Neil, Professor Emeritus of Anthropology Behavioural Sciences Department at Palomar College has claimed that modern East Asians commonly have shovel-shaped incisors, something typically seen in *H. erectus* fossils but not seen often among Europeans or Africans. He refers to opinions shared by Alan Thorne of the Australian National University (now deceased) that Aboriginal Australian peoples share skeletal and dental traits with the oldest known inhabitants of Indonesia. His conclusion is that the evidence supports the hypothesis that no replacement of local humans by African migrants ever occurred.

Highly distinguished Chinese paleoanthropologist, Professor Wu Xinzhi, stated in a recent interview, "There is overwhelming evidence from fossil records that China was populated with humans before the arrival of African settlers". It is easy to think that such sentiment is merely hype being pushed by proud Chinese nationalists, but keep in mind that certain transitional fossils in the region are already well accepted by international scientists. One thing nationalism is doing is encouraging governments to support more paleoanthropological research in Asia, which is no bad thing. It is becoming difficult to disagree with some of the bold statements issued by evolutionary researchers working in the Asian region.

The evidence is now strongly supportive of archaic *H. sapiens* moving around in East Asia alongside the early Neanderthals known as *H. heidelbergensis*. The timing of their presence in China fits the requirements for our search. It is tempting to think that we are at the end of our journey, that we have found the region from which the three branches of anatomically modern humans arose. We can begin to imagine groups of early *H. heidelbergensis* moving west from their East Asian homeland, entering both Europe and Africa, later followed by archaic *H. sapiens* that would interbreed with their distant cousins during the migration.

There is just one major problem with the picture arising from this data. Professor Alan Cooper, University of Adelaide, points out that mainland Asia lacks the genetic signal for a Denisovan interbreeding event. The trace here is so faint that we just can't view this as a viable possibility for territory in which these hominins lived in large numbers, regularly encountering populations of Neanderthals and *Homo sapiens*. The genetic profile of modern East Asian populations offers trace levels of Denisovan DNA, most commonly around 0.1% of the genome, ranging only as high as 1% among a small number of individuals.

There is only one region on Earth where we find the genetic signature of long-term interbreeding between Denisovans, Neanderthals and fully modern humans. Professor Cooper suggests that the genetic data places interbreeding between Denisovans and modern humans somewhere east of Wallace's Line, the natural geological barrier between Asian and Australasian flora and fauna. Even though Professor

Cooper's positioning requires Denisovans to have been on the wrong side of a supposedly impenetrable oceanic barrier, beyond the limits of South East Asia, that is what the evidence supports. Before we examine the Denisovan interbreeding data in more detail, we first need to understand how archaic hominin populations crossed from South East Asia into Australasia.

Our discoveries in Central and Eastern Asia have continued to undermine the, already severely weakened, Out of Africa theory. We have witnessed the relocation of ancestral Neanderthals from Europe to Central and Eastern Asia, the geographic location for interbreeding between them, and modern humans shifting with this. During the examination of archaeological evidence in East Asia, we uncovered fossils of early Neanderthals situated close to those of early *Homo sapiens*, the second of these populations being resident long before they are supposed to have migrated from East Africa. These findings support a model in which *Homo sapiens* migrated west from China, entering the Levant and Africa before finally one group crossed the Eurasian border and entered Europe. What remains questionable is whether this journey started in China or further afield. The fact that Denisovan DNA shows up in the populations of East Asia confirms for us that we are approaching the founder population shared by all three anatomically advanced *Homo* forms. We already know that our target group lived between 900,000 to 700,000 years ago and we have now narrowed down their location. By an incredible twist of fate, we now find ourselves on route to the land that gave birth to modern human evolutionary theories, once playing a central role in an *Out of Asia Theory* for human evolution. We require some time in Java.

Chapter 4 - *Homo erectus* becomes the First Orator, Sailor and Artist

It may come as a surprise to learn that in 1891, Eugène Dubois, a Dutch paleoanthropologist, discovered the first known fossil remains of a *Homo erectus* on the banks of the Solo River, in Java, Indonesia. He soon became one of the principal founders of the Out of Asia Theory for human evolution, which remained a hugely popular model in the scientific community right up until the mid-20th century. A German evolutionary biologist, Ernst Haeckel, took the theory even further, claiming that all hominins had evolved from a primate species resident in Southern Asia. This unusual approach to the subject involved humans evolving on a landmass out in the Indian ocean that he called Lemuria. This Lemurian gateway, existing between Asia and Africa, supposedly became submerged during a geological catastrophe. We may not be very concerned with the wilder aspects of these theories, but certainly, we should keep an open mind about what was once the leading evolutionary theory in the world: The Out of Asia theory.

In recent years, I have found myself reading about incredible discoveries made in South East Asia, none more exciting than the ancient pyramid site of Gunung Padang situated in eastern Java, considered to be the largest megalithic complex in the entire region. At Gunung Padang, it is not so much the size that has stunned researchers but rather the results of the archaeological dating process. If the evidence is correct, Gunung Padang is up to 20,000 years old, the oldest human-engineered structure ever uncovered. That date connects us to the last remnants of a mysterious hominin population living on Java known as *Homo erectus soloensis*, a subspecies of hominin displaying morphology that is transitional between *H. erectus* and *H. sapiens*. Despite dense bones, thick brow ridges and flattened skull profiles, these people had limb bones indistinguishable from those of modern humans. The *H. erectus soloensis* brain was almost as large as ours, ranging from 1150cc to 1300cc. We must be heading in the right direction geographically, but we need to go much further back in time.

Fossil finds have long proven that Java boasted a well established *H. erectus* population by 1.6 Mya, but the activities of interest to us occurred about 600,000 years after that time. We now find ourselves back in the era associated with the global population collapse and close to the period in which the three identified branches of anatomically advanced hominins diverged away from their common ancestor. The event we have come to investigate is the first sea journey, beginning with a crossing from Java to the Indonesian island of Flores.

During the 1950s and 1960s one Father Theodor Verhoeven, a Dutch Catholic priest based in Java, discovered multiple ancient stone tool sites. The scientific community steadfastly ignored Father Verhoeven's suggestion that the tools might be evidence of *Homo erectus* on Java some 750,000 years ago. We now know that these tools do prove an early presence of hominins migrated to Flores. New dating in 2010 headed by Adam Brumm, an archaeologist at the University of Wollongong in Australia, pushed the date back to around 1 Mya. The presence of undisturbed volcanic material covering some of the tools allowed for more reliable dating.

In 2003 the Indonesian island of Flores came to the attention of the public as the world's media excitedly reported the discovery of a strange species of small humans nicknamed *Hobbits*. The ancient hominins of Flores, correctly known as *Homo floresiensis*, were incredibly small in stature, standing just a little over one metre in height. Dating of the *H. floresiensis* bones discovered at the Liang Bua Cave suggested an initial arrival onto Flores around 190 Kya, with the community existing until 17 Kya. There was naturally lots of suspicion on a connection between these diminutive hominins and the significantly older stone tools already found on the island; the problem was the lack of sufficiently old fossils.

It was only in 2016 that new evidence emerged which offered a link between the ancient tools and the more recent fossils. Several fossils were found, including a jawbone and some teeth, representing at least three individuals. The fossils came from human adults yet were proportionately tiny, dating to around 700 Kya. The size match left little doubt that these were ancestors of the later *H. floresiensis* population inhabiting the Liang Bua cave site.

There were conflicting interpretations of the available data.

Many researchers suspected that a group of Javan *H. erectus* had become isolated on Flores, and then had undergone island dwarfism, as indeed had another species stranded on the small landmass, pygmy elephants. Evolutionary biologists struggled with the idea that humans could experience such a drastic size reduction in just 300,000 years, decreasing from around 1.5 metres in height to just over 1 metre. This was unprecedented in all known biological records. This theory of shrinking *Homo erectus* was further eroded when it became clear that the older fossils represented individuals that were 20% smaller than the later form. Further detailed analysis revealed morphological features associated with hominins more archaic than the *H. erectus* on Java, causing many scientists to voice the opinion that *Homo floresiensis* had evolved from an already diminutive hominin, perhaps an *Australopithecus* or a very early *Homo* species. The Australian Museum noted that the physical features best fitted a scenario in which hobbit people shared a common ancestor with Asian *H. erectus* but did not descend directly from *H. erectus*. Could it be that a hominin species more archaic than *H. erectus* had migrated into Indonesia around 2 million years ago?

Two scientific researchers, Bernard Wood and Alan Turner, have offered the interesting theory that early *Homo* groups could have left Africa around 2 million years ago, evolved further in Asia, and then regressed to Africa. These scientists speculated that Asian *H. erectus* could have amigrated back into Africa and formed the root population for so-called *H. ergaster*. Wood and Turner's model allows for gene flow in both directions, with outwards and inwards migrations by a single *Homo* species inhabiting various geographical regions. This two-way movement theory fits well with the findings from Flores and the Dmanisi site. Both suggest a single evolving *Homo* species with high levels of correspondence across geographic regions.

African hominins, those currently assigned to the *Australopithecus* genus, certainly had the small stature and average brain sizes we would expect of hobbit ancestors. *Homo floresiensis* had just 380cc of grey matter, considerably less than the 1000cc evidenced in *Homo erectus* fossils from the corresponding period. Flores was providing proof of a separate evolution of *Homo Australopithecus,* (or perhaps Archaic *Homo erectus*) involving a population living well outside of Africa. If hominins did leave Africa considerably earlier than fossils have so far revealed, it might be that Asian *Homo erectus* also descended from unidentified first residents of Asia.

Homo floresiensis persisted on Flores right up until as recently as 50 Kya, leaving some scientists to ponder whether there was modern human involvement in their demise. The debate over how *H. floresiensis* became extinct was a central theme in at least one paper published in *Nature*. There often seems to be an assumption bias that favours the idea that we modern humans always killed off other human species we encountered, but in fact, our genome suggests that we are more likely to interbreed with such populations until we fully absorb them.

In Southern China, hominin remains discovered at the Red Deer Cave archaeological site have proven that hominins displaying primitive morphology had lived alongside modern humans until as recently as 12,000 years ago. Modern humans would have certainly encountered the Red Deer Cave people, living well inside their timeline and in the same region as them, and yet we seem to have felt no need to massacre them. Evidence gathered on site, mostly examined by Associate Professor Darren Curnoe of the University of New South Wales and Professor Ji Xueping of the Yunnan Institute of Cultural Relics, gives us a picture of a peculiar hominin. Red Deer Cave hominins weighed around 50 kilos and exhibited a mixture of both modern and archaic features – including a thigh bone much the same as those of hominins living 1.5 million years ago. It might yet transpire that this hominin shares ancient ancestry with the Hobbit people of Flores.

The archaeology on Flores points towards a crucial development in the hominin timeline that is far more exciting than simply one new physical form to consider. The island itself sits right on the Wallace Line, the natural border dividing flora and fauna of Asia from that of Australasia. Deep sea trenches in this region ensure that, even with the rise and fall of the ocean during several climate shifts over large periods of time, these islands are always cut off by oceanic waters. Between the islands of Indonesia flow prevailing southerly currents, transporting huge volumes of water from the Pacific into the Indian Ocean. Migration beyond Bali, on to Flores, required a minimum of ten challenging sea crossings ranging up to 100km. The only rational way that hominins could have reached Flores was by developing a new technological advance: watercraft. With the first ever sailing we are forced to recognise an enormously significant leap in cognition.

Palaeoanthropologists have long claimed that only fully modern humans could build sea vessels and successfully set sail to new locations. The traditional consensus theory on human origins includes the claim that sailing first appeared around 60,000 years ago when a group of modern humans migrated across the Wallace Line into Australia. The voyage to Flores represents both the first use of watercraft and a much earlier instance of hominins successfully penetrating the Wallace Line, a bitter pill for academics to swallow; many still refuse to even look at the matter seriously. As incredulous as it is, serious researchers are suggesting that early hominin populations arrived on Flores by clutching to debris carried away after a tsunami wave. It produces quite a mind-boggling picture to imagine an entire founder population clinging to driftwood, swept east across powerful southerly currents, not once but ten times.

Watercraft represents the only way in which hominins could arrive on Flores. The mystery is whether these first sailors were *Homo erectus* or some other archaic form. For those who feel that they would require further evidence of archaic hominins using watercraft, we can consider some additional evidence before moving forwards.

The second-oldest sea journey identified so far was to the

Socotra, an isolated island far out in the Gulf of Arabia. Socotra is situated 240km from the Horn of Africa and 385km from Yemen. Either starting point makes for an incredibly long and arduous voyage by primitive watercraft. The distance to Socotra makes it an impossibility for humans to swim across or float on a log. Russian researchers investigating the island in 2008 uncovered several stone tools that dated to at least 500,000 years in age, possibly closer to one million years. The investigation of the site was not well publicised, though thankfully it did feature in the magazine *Science Illustrated* (2012).

Moving away from the Gulf of Arabia, we find further evidence of ancient hominin sailors on the Mediterranean island of Crete in the form of over thirty stone axes (and other tools) spread across the island. The high number of artefacts dispels suggestions that one or two accidental castaways might have reached the island. Reaching Crete by watercraft would have required a series of sea crossings. Discussions of the finds have featured in the *National Geographic* magazine, with involved geologists dating the oldest terraces associated with axes at no less than 130,000 years of age.

The last archaeological site for us to consider in association with ancient sea voyages is perhaps the most relevant to our wider investigation. This time we are back in Indonesia. Archaeologists on Sulawesi, another island positioned east of the Wallace Line, have recovered stone tools. The stone tools on Sulawesi have been suggested to be around 200 Kya, although there have been speculations that a few of the artefacts might be up to one million years old. Due to the direction of the currents in the region, it makes a great deal more sense to imagine Flores populated by voyagers from Sulawesi, rather than from Java. Could our sailors have been moving from west to east, rather than east to west?

As to whether the first sailors in South East Asia were *H. floresiensis*, *H. erectus*, or yet another different hominin, and what their direction of travel was, we can only speculate. As to what type of watercraft they used, we can likely look to modern humans for possible options. On his personal website, the distinguished engineer Jean Vaucher (former professor at Montreal University) has written on this very subject under the title *History of Ships, Prehistoric Craft*. Vaucher readily discusses some of the simple watercraft capable of reaching Flores; these include bamboo rafts, reed boats, dugout canoes and coracles. The vessels described by Vaucher are easy enough to construct with only primitive stone tools; his favourite is the bamboo raft. The material required to build a bamboo raft was always readily available in Indonesia. Although simple to assemble, these can potentially carry scores of people on lengthy journeys. It is by no means an immense stretch for us to imagine early hominins lashing a few bamboos together with tough vines.

Although constructing a large bamboo raft might be relatively straightforward, organising a migration across water necessitates archaic humans communicating a radical new idea amongst a group. Unlike simply walking over the next hill to pastures new, we are talking about a much more complicated migration scenario. Our boat builders had to do more than just constructing a vessel; they needed to convince a significant number of men and women of breeding age to board a flimsy looking vehicle and set sail over potentially dangerous waters, with a mysterious new home somewhere in the distance. With population migrations by sea comes a requirement for highly complex social interactions. It is hard to imagine such a journey can take place due to a single man or woman running around gesticulating wildly and pointing animatedly at the open sea. Does such a scenario sound as insufficient to you as it does to me?

If you wish to communicate a complex idea, the ideal tool to possess is a spoken language. We will never have a definitive date for the first word uttered on this planet – the use of meaningful sounds goes back millions of years, and they are a feature of various species. We can at least explore the possibility, or impossibility, of spoken language having existed among archaic hominins. There are several anatomical components involved in producing speech, and these indicate at least whether an ancient human possessed the potential for articulated language.

There exists strong anatomical support for speech capabilities among early hominins, especially found in the work of Anya Luke-Killam and in a highly relevant paper by Antonio Benítez-Burraco1 & Lluís Barceló-Coblijn. The work of Luke-Killam refers to research carried out previously by Wynn et al. in which researchers described the brain of *Homo erectus* as having possessed a similar structure to that of fully modern humans, quite unlike that of apes. Wynn interprets the evidence provided by fossil brains as suggestive that *"Homo erectus* may have had a vocal language". There are two distinct regions of the brain directly associated with speech, from a neurological perspective: the lower left frontal lobe and the left parietal lobe. Fossil brains, or rather we should say casts of the internal area of fossilised skulls (*endocasts*), suggest to us that the key areas of the brain such as the Broca's area and Wernicke's area are very human-like in *H. erectus*. The available evidence suggests that hominins living on Earth one million years ago in general, and most certainly Asian *H. erectus*, would have been anatomically ready to produce speech.

Keep in mind that an ancient system of speech need not be as complex as modern languages. The English language with its 170,000 words would almost certainly be too much for any ancient hominins to replicate. There are forms of language even today that require little-complicated pronunciation and far fewer words than are in English. The *Khoisan* language system (associated with several distinct Khoi and San cultural groups in Sub-Sharan Africa) incorporates a mixture of words, tongue clicks, whistling sounds, replications of animal and bird calls, in combination with physical gestures and body language. Such communication systems as this clicking language allow for efficient complexity without the need for lots of difficult words.

The remnants of another click-based language, currently on the edge of extinction, is found amongst the Lardil and Yangkaal people, traditional inhabitants of islands of the Gulf of Carpentaria. *Damin* existed in historical times as a language used solely for ceremonial purposes, only known to a few initiated men, but it seems likely that it was once a commonly used system. Tantalising evidence also suggests that a third click-based language existed among the indigenous people of Tierra del Fuego, just off the tip of southern Argentina. The people of Tierra del Fuego Island are remnants of an incredibly ancient population, possibly the last descendants of the very first migration into the American continent. The only clues remain hidden among the letters of Charles Darwin as recorded in the book *Whitley, Life and Letters of Charles Darwin*. Darwin wrote of the language spoken on Tierra del Fuego that it "scarcely deserves to be called articulate" adding that Captain Cook had "compared it to a man clearing his throat, but certainly no European ever cleared his throat with so many hoarse, guttural and clicking sounds." The final clues from Darwin are in his description of a local man communicating from the shore: "Standing on a rock he uttered tones and made gesticulations than which, the cries of domestic animals are far more intelligible."

The obvious common thread between these three click-based languages is that they are each associated with ancient human populations. In fact, the oldest human societies on Earth are the Australian Aboriginals and Sub-Saharan Africans. It is important to mention that I am currently involved with ongoing investigations into the ancestry of the people of Tierra del Fuego; evidence has already emerged that suggests to me that they originated from a pre-Clovis migration into the Americas involving prehistoric Australasians. There is a good reason to suspect that our ancient sailors could produce clicks, whistles and hand gestures, at the very least, and reason to speculate that they could have mastered a few basic words.

Before leaving the topic of the first communication methods, there is one final puzzle piece from Indonesia that we must consider even though it comes from a slightly more recent period. Recent archaeological efforts on Java uncovered the very first known use of modified sea shells, some having sharpened edges making them efficient for both cutting and scraping. The real surprise was that around 500,000 years ago the hominins on Java started engraving geometric patterns onto these shells, representing the very earliest abstract art known on Earth. Scientists involved with the Java shell discoveries wrote in their paper that *the manufacture of geometric engravings is generally interpreted as indicative of modern cognition and behaviour.*

Modern cognition and behaviour relate to everything we think of when we consider how modern humans differ from other Earth creatures: dancing, speaking, singing, writing, loving, and philosophical thought. The implications of hominins exhibiting advanced cognition and behaviour 500,000 years ago are nothing short of profound. The geometric pattern seen in the Javanese shells can be considered as symbolic communication in the same way that art today conveys messages to the one regarding it. A growing number of ancient stones exhibiting engraved patterns similar to the Java shell exist in the Australasian region (not yet featured in any published studies) and a crosshatch pattern engraved into a shell found near Pilbara is dated to 30 Kya. The zig-zag pattern on the shell is almost identical to patterns seen engraved on valuable artefacts from South Africa known as the *Blombos stones*, dated to around 70-80 Kya. The Blombos engravings are considered crucial evidence for the first emergence of higher thinking in *H. sapiens*. It now seems the prototype to the Blombos stones had been engraved in South East Asia several hundreds of thousands of years previously.

The finds on Flores and Sulawesi dispel the myth that ancient humans could not cross the Wallace Line. Top academics have already informed us of a strong link between the evidence of *H. sapiens* interbreeding with Denisovans and the lands east of the Wallace Line. Step-by-step, our investigation has narrowed down the possible location of the ancestors we shared with Neanderthals and Denisovans. The only region remaining to consider is that vast landmass to the east of the Wallace Line. We find ourselves gazing towards the coastline of what is today Australia.

Chapter 5 – The Australasian Homeland of the Denisovans

When we look at a modern map of the world, we forget that it did not always look the way it does today. Our planet cycles through ongoing periods of rising and falling temperatures. As temperatures decrease water freezes into vast glaciers, sea levels drop considerably, and enormous areas of land emerge from the oceans. Nowhere is this change more observable than in South East Asia and Australasia. During times of low sea levels the landmass of South East Asia was far more extensive than now, the Java Sea vanished, and the Gulf of Thailand diminished to a large inland lake. Even the South China Sea receded to such an extent that it became possible to walk from Vietnam to the Philippines. This changed landscape is known as the Sunda shelf, encompassing 1.85 million km² of land. Across a residual, short, stretch of water, we find the Sahul shelf. This vast landmass incorporates all of Australia, New Guinea and Tasmania, as well as significantly extending the entirety of the Australian coast. When global temperatures increase glaciers melt, and the sea levels rise; the ocean swallows the land once again. This cycle plays a significant role in our current investigation.

Palaeoclimatology is the scientific discipline dedicated to studying changes in Earth's climate over geologic epochs. Thanks to the scientists who have dedicated themselves to this branch of scientific inquiry, we now have records of climatic shifts going back millions of years. David Lappi, a geologist from Alaska, explains to us that over the last 1.25 million years until the present thirteen ice ages occurred on an approximately 100,000 year-long cycle. Each glacial period is around 90,000 years long with warm interglacial periods of about 10,000 years. In 2005 climate scientists Lisiecki and Raymo found evidence in sediment cores representing at least two major temperature drops, falling to around 8°C lower than today's temperatures, during the period from 1,000,000 years ago until 800,000 years before the present. We now know, from the analysis of ice core related data, that global temperatures were 9°C lower 800,000 years before the present.

Palaeoclimatological data enables us to understand the geographical situation encountered by hominins living on the Indonesian islands 1,000,000 years ago. Periodic sea level drops allowed their sailors easier exploration of local islands using only simple water craft. In the coldest periods, sea levels were approximately 120 metres lower than today; the distance from the Australian coast to Timor (the most easterly Indonesian Island) dropped to 90km. Any sea vessel capable of carrying hominins from mainland Sahul across the mighty currents of the Wallace Line to Flores and Sulawesi could also take them to Timor and Australia.

In 2007 Mike Morwood, one of the paleoanthropologists in the Flores research team, made a prediction that further species of hominin would be shown to have lived on the islands of Timor and Sulawesi. Sulawesi has since provided stone tools dating back to 200 Kya. Similar devices have been found on Timor but are not yet accurately dated. One of the stone artefacts on Timor found by Robert G. Bednarik, a scientific researcher, was discovered in association with a middle Pleistocene geological formation, suggesting that it had been manufactured somewhere between 781 to 126 Kya. In 2006 Sue O'Connor, an archaeologist at the Australian National University in Canberra uncovered a cave site in East Timor replete with the remains of deep sea-dwelling shark and tuna alongside at least one fishing hook. O'Conner set a minimum date for inhabitation of the site at 42 Kya, the limit of the carbon dating methods used. However, the cave was almost certainly in use for far longer than that. Stone tools uncovered in the cave are like those of the *Homo floresiensis* hominins that vanished 50,000 years ago.

Hominins living on Eastern Timor could have seen the western edges of the Sahul landmass from any elevation above 1500m. Mount Ramelau, Timor's highest point, reaches an elevation of around 3000m above sea level and provides unbroken views for 200km. It is of course equally probable that a fishing fleet of watercraft observed the coast of Sahul on the horizon.

Professor Alan Cooper, University of Adelaide, has recently voiced his opinion that any hominins able to cross Wallace's Line during a period of low sea levels should then have reached Australia. As he said in his conclusion, "If you cross Wallace's Line you've done all the hard work". The difficulty in proving such an arrival event would be the fact that a 120m sea level rise claimed around 1,551,600 square km of the Sahul Shelf. 310,800 square kilometres of the sunken area were precisely where any voyagers would have landed and made their initial settlements.

Without hope of direct physical archaeology dating back to the time of the early settlers on Flores, we are to be forced to return to genetic research projects. With some luck, we can track down evidence of the ancient migrations we have been following and ideally pick up the trail of the mysterious Denisovan population. One critical study carried out by both Harvard Medical School and UCLA created a world map of ancient DNA. During their process, the academic teams discovered that the populations of Original Australians and Papuans evidenced up to 5% Denisovan DNA, alongside the usual levels of Neanderthal DNA (between 2-3%). The study confirmed that only trace amounts of Denisovan DNA remained present in South Asia and of East Asia, close to 0.1%. Natural selection helped modern humans to retain certain positive traits from Denisovans, including a boosted immune system and adaptations to a sense of smell. Negative traits inherited from Denisovans, including infertility-causing genes, were removed by adverse selection. Genes that impair fertility come as a consequence of interbreeding between populations en route to becoming separate species. A key finding of this study was that the interbreeding event had happened approximately 40 Kya.

Professor Alan Cooper has already pointed out to us that mainland Asia lacks any signature for an ancient Denisovan-human interbreeding event and that the only area where it is present is east of the Wallace line on Sahul. Professor Cooper suggests Denisovans sailed across the Wallace Line and entered Australasia around 100 Kya, followed later by modern humans around 50 Kya. The two groups apparently crossed paths and interbred around 40 Kya. Cooper seems completely unaware of hominin populations sailing across the Wallace Line around one million years ago. At least, he makes no mention of them. Exactly why every human group that crosses the Wallace Line can easily make it to Australia apart from the hominins sailing Indonesian seas one million years ago is a matter never clarified. Professor Cooper does point out one major puzzle arising from the UCLA mapping initiative: "They say there is only one invasion [of Sahul] and that happens around 50,000 years ago, but then they go on to say Aboriginal Australians have genetically mixed with Denisovans at 44,000 years."

We can see here that the consensus theory on the populating of Australasia just can't support these findings. As usual, the study assumes Australia was entered for the first time 50,000 years ago by a single population of modern humans. Genetic data used for dating Australasian populations always contains assumption bias with respect to an entry date, calling into question the validity of these studies – something we will discuss in more detail later. Here we see that the research findings show two evolutionarily distinct hominin populations interbreeding on the continent somewhere between 40 – 44 Kya. This acknowledged interbreeding event presents an insurmountable and obvious flaw in the current origin theory; we see here evidence that directly contradicts stated assumptions.

We can perhaps assist Cooper and the UCLA team by pointing out to them that there is no indication of any large Denisovan populations resident anywhere west of the Wallace Line. Denisovans appear to have had no territory from which to have sailed into Australasia; this would seem to limit their emergence to that very continent. The assumption that they *must* have sailed into Australasia relatively recently is quite simply a result of assumption bias, an attempt to patch up the crumbling Out of Africa Model with its single human migration from Africa 70 Kya. We English would refer to this desperate grasping as 'clutching at straws'. Australasia is the only possible homeland for the Denisovans; the DNA evidence firmly places their population centre there and only there.

Further genetic studies only seem to raise further scientific conundrums. One interesting study carried out by an international team, headed up by Professor Eske Willerslev of the University of Cambridge, specifically focussed on Original Australian and Papuan groups. The results of this study, undertaken with the collaboration of elders and leaders of various Indigenous communities, were published in *Nature*. The team sequenced the genome of 83 Original Australians from the Pama-Nyungan-speaking language group, which covers 90 percent of the continent, as well as including 25 Highland Papuans.

The study of Original Australasians revealed two highly problematic insights: evidence of a 4th and unidentified archaic hominin population and data suggestive of modern humans being already widespread across the entire continent by as early as 45,000 years ago. Speaking to the UK-based *Telegraph Newspaper*, Professor Eske Willerslev offered a few thoughts on this mysterious archaic hominin species: "We don't know who these people were, but they were a distant relative of Denisovans, and the Papuan/Australian ancestors probably encountered them close to Sahul."

I think we can narrow down the location from the vague suggestion that modern humans "probably encountered them close to Sahul" to a much more precise statement, that modern humans definitively met Denisovans on Sahul. We find strong genetic evidence of these hominins only in the one region, and there they provided as much as 4% of the modern Australasian genome.

The academics are slowly heading towards my own logical deduction that Denisovans lived in New Guinea, as can be seen in a recent utterance from Professor Richard Roberts, Director of the Centre for Archaeological Science at the University of Wollongong: "The existence of Denisovan DNA in the Aboriginal Australian's genome indicates that the original dispersing population of *Homo sapiens* must have encountered resident Denisovans en route to Australia, possibly in New Guinea. So we now have the intriguing possibility that the island chains leading to Australia were home to the last surviving members of *Homo erectus* on Java, 'hobbits' (*Homo floresiensis*) on Flores, and Denisovans in New Guinea – and that some or all of these were met by the ancestors of the Aboriginal Australian whose hair was sequenced in this study."

Academics have continuously suggested that a small influx of modern humans rapidly expanded into a large population, inhabiting much of the vast Australasian continent within a mere 5000 years of the proposed arrival date. How a handful of migrants clambered out of their boats on the North Australian shore and a few thousand years later dominated a vast continent, well, it is a difficult claim to process. Even the research scientists struggled to make sense of their findings, offering a few possible explanations, all of which stretch credibility. There is no possible way that we can reconcile these results with the current consensus theory in which a tiny migrant group populated an entire continent in a short period. We have to remember that we also have interbreeding occurring between *Homo sapiens* and two other archaic human species that have no business existing in this region, according to the official version of early migrations.

For a supposedly tiny and isolated human population, Original Australasians seem to have managed an incredible level of interbreeding with other hominins. The average level of archaic hominin DNA amongst the population is apparently around 10%, identified so far, including contributions from at least three distinct human forms. This level of archaic hominin DNA makes these people unique on the planet. It will come as no surprise to me if we go on to detect yet more unknown hominins represented in their DNA profile, though I am sure leading minds in this field will be utterly baffled.

For distinct hominin groups to interbreed on any significant scale, often enough that their DNA remains present at high levels tens of thousands of years later, they must spend significant periods of time alongside each other. Two very different groups of humans (culturally and physically) passing each other briefly during migration events are unlikely to do much intermingling. We have already seen that in areas beyond Australasia, regions where Denisovans presumably passed through, their DNA did not survive at significant levels in the modern human populations. Even in the area surrounding the Altai cave site, we find no trace of Denisovan genetics among modern populations. We can, therefore, infer that significant levels of interbreeding occurred within the boundaries of Sahul, and over many generations. High levels of interbreeding suggest that relatively large populations of *Homo sapiens* and Denisovans lived alongside each other in Australasia. Just as Sergi Castellano, an evolutionary biologist at the Max Planck Institute for Evolutionary Anthropology, says on this matter: "One would think that mixing has occurred multiple times for a long time." We start to see a picture forming here, one in which three or more significantly-sized populations of advanced *Homo* forms controlled separate segments of the same continent (Sahul), resulting in inevitable intermingling.

At the end of 2016 results were published from a major genomic investigation, *The First Genomic Study of Indigenous Australia,* run by Griffith University's Australian Research Centre for Human Evolution, and the University of Copenhagen. The team was astonished to discover that by 37

Kya the population inhabiting the region of Papua New Guinea had already separated away from other populations on the continent. This was at a time when there was no geological separation – the rise in sea levels only separated the landmass around 10 Kya. The researchers involved failed to make the connection between the Denisovan interbreeding event already revealed in Papua around 44 Kya and the isolation of humans in that region that followed within a few thousand years. The scientists did suggest a restriction on intermarriage might have been the cause of the separation but missed the glaring reason why intermarriage between these regions would have been made taboo. Though this next suggestion is largely speculation, it may even be that some of the inhabitants of the Denisovan cave in Siberia were exiled offspring of individuals that broke the interbreeding taboo. Analysis of the 50,000-year-old female Denisovan discovered in the Siberian cave revealed that her ancestors had interbred with the ancestors of modern New Guinean's, necessarily long before the recognised interbreeding event dated to 44 Kya.

If we view the separation between north-east and south-west sections of Sahul with the understanding that a Denisovan population lived within defined territorial boundaries in the north-east, we can immediately clear the fog of intellectual confusion. The modern humans that merged with Denisovans 44 Kya inherited the same intermarriage taboo that already existed between modern human tribes and Denisovans. It should be noted that the most significant traces of Denisovan DNA have so far only been found in New Guinea and nearby parts of North Australia, precisely where I suggest Denisovan territory extended to on Sahul. The Griffiths University study also noted that the divergence between Aboriginal Australian people of north east and south west Australia makes these populations more genetically different than are Native Americans from Siberians.

The suggestion that a fourth human species was present and interbreeding with the ancestors of modern Papuans and Aboriginal Australasians has been confirmed yet again in October 2016. Ryan Bohlender, a statistical geneticist at the University of Texas, had been conducting a deep analysis of archaic hominin DNA present in the genome of modern human populations. Initially, Bohlender's team confirmed the usual levels of Neanderthal DNA among Original Australasians at 2.8%, close to the figure represented in all non-African populations. Bohlender's findings for Denisovan DNA were significantly lower than in past studies, with East Asians attributed a level of 0.1% and Melanesians 1.11%. The reason for the discrepancy in the levels of Denisovan DNA among Melanesians was the revelations that some of the material previously identified as Denisovan turned out to be from yet another, unidentified, hominin. We can reasonably assume that this fourth human relative was the same one as previously detected by Professor Eske Willerslev. Bohlender's study awaits full analysis by the research community, but he is confident that either there is another human lineage or else the current research is failing to understand the relationships between modern humans and the other hominin populations.

There should be no surprise to find that Papuans carry remnants of additional lineages closely related to Denisovans. I predict that many more novel lineages will be uncovered in the same geographic region. The fossil fragments from Siberia revealed high diversity among the small number of Denisovans present, as well as an unknown hominin in their ancestry. All of this suggests that there was once an enormous population of Denisovans (and close relatives). The population of Denisovans living in north-east Sahul must have been little more than a small remnant of this once mighty race. We know this because it was easily absorbed into a small population of modern humans with the older group's DNA becoming no more than a trace remnant. Further support for the claim that New Guinea was home to the remnants of a much larger Denisovan population can be found in the local languages. The total number of languages spoken on our planet is close to 7000. Of these, 820 are considered Papuan languages. This is the highest level of language diversity on the planet. With a population of just over 7 million, this gives us one language per 8563 people. The languages themselves are grouped into as many as 30 language families, and many of these languages are totally unrelated to any others.

It starts to seem that some of the greatest minds on Earth are unable to see the picture right in front of their faces. We can at least understand why evolutionary scientists are suffering from signs of cognitive dissonance; they have long laboured under the delusion that data should always fit into the Out of Africa paradigm. The field of evolutionary science boasts many incredible researchers, genius minds, yet they are unable to move beyond the popular assumption that a single ancestral population exited Africa around 70 Kya and made a single entry into Australasia between 60 - 50 Kya. All too often, when OoAT scientists encounter square-shaped data, they shave the corners off and fit it into the same old, round hole. Even as researchers publicly admit that Aboriginal Australians are the oldest culture on Earth, they continue to attempt then to place abstract limits on how ancient that culture can be.

We have now seen that it is impossible to defend a single entry into Australasia (Sahul) by modern African humans 50,000 years ago, followed by isolation. Australasia is transpiring to be the most exciting region of the planet when it comes to human origins and important events in our evolutionary story. We will soon discover that this ancient land was much more than a home for the Denisovans or a refuge for wandering modern humans. We have finally reached 'site zero' for the most important event in the entire human story.

Chapter 6 - Evolutionary Emergence of *Homo sapiens* in Australasia

The various populations indigenous to Australasia spent most of the 20ᵗʰ century as a mere footnote in human history. It was long popular to think of this region as the last place anybody migrated into and a land of primitive people frozen in time, achieving nothing of note. There were many scientists of past eras that even felt Original Australasians represented a devolution back to animal status. This negative perspective on the indigenous populations of the continent grew out of the picture painted by the first English invaders; they were keen to portray the victims of their genocidal incursion as being little more than apes. As we all know, killing animals raises far fewer criticisms than does the slaughtering of fellow humans. The strategy used in Australasia was one used by most colonial powers, an effective form of propaganda known as *dehumanisation*. Dehumanised populations can suffer almost any horror without the protection of a widespread societal backlash. It is a terrible fact of history that the people we know as Australian aboriginals existed in the same legal classification as crocodiles, koalas and kangaroos right up until 1967!

The first European perspective of Original Australians was that they were recently arrived at their continent, perhaps present for less than 20,000 years, and then remained stuck in the stone-age. This set of negative assumptions meant that evolutionary researchers and archaeologists alike took little interest in exploring Australasia. It is only in recent decades that paleoanthropological studies have begun across the region in any depth of seriousness. As we have seen in the previous chapter, and as we will continue to see, the research that has happened is already turning the entire scientific paradigm onto its head. To fully appreciate how profoundly wrong the OoAT is in respect to Australian Aboriginal peoples we need to consider the work of two of the most prominent Out of Africa proponents, Professor Alan Wilson and Professor Rebecca Cann.

Professors Wilson and Cann co-authored a seminal paper on the African origin hypothesis, *The Recent African Genesis of Humans* (1992). The conclusion they offered to the world was that a single African woman was at the root of our mtDNA family tree, or rather, that is if all their stated assumptions were validated, then our lineage was probably African and approximately 200,000 years old. Some of the key assumptions don't make sense considering recent discoveries.

Being two of the most highly respected academic stars in the field of evolutionary biology, we can hardly turn to scientists of a higher pedigree. At the time of publishing their most famous paper, Wilson and Cann shared in the commonly accepted view that Australia had been populated by a group of Africans 40 Kya as they entered South East Asia around 50 Kya. These fundamental beliefs have also since been shown to be incorrect.

It is not well known that these two giant figures of the Out of Africa paradigm were not always so sure of the model they went on to make famous. In the early 1980's Cann had carried out cutting-edge genetic research work involving the mitochondrial material (mtDNA) of several population groups. This project, involving samples from 112 individuals from diverse geographic regions, had uncovered a wide range of anomalies that are not well publicised but which hold immense importance. Some of the anomalies call into question the existing paradigm in human evolution. One of the first of these findings was the strange matter of human variation or rather a lack of it. While it is normal to find a difference of around 1.5% between the genetic material of any two mammals of the same species (including chimps), it transpired that any two humans differed by just 0.4%. The lack of variation immediately suggested that humans were in some way a very recently emerged species, significantly more recent than our closest primate relatives.

The lack of variation among humans was not the most notable finding from Cann's study. Until her investigation, the human family tree had been drawn up from fossil finds, and the examination of blood proteins, with the major groupings of humanity (then considered to be; black, white, and Australian-Oriental) positioned 100 Kya, with the split between Australian and Asian people calculated to have occurred 40Kya. We should keep in mind here that in 1982 the oldest known fossils of an entirely modern human being were dated to just 39 Kya. Through the process of examining mtDNA gathered from more than one hundred individuals, of diverse racial backgrounds, Cann ended up with a very different family tree to the protein-based model. In the revised family tree, all races were tightly grouped together, the exception being a small number of Australian Aboriginals and Asians that differed significantly from the bulk of humanity, but were very placed very close to each other.

The strangeness in Cann's results continued as she calculated the timing of splits between groups. The divergent Aboriginal Australians had split away around 400 Kya, the anomalous Asians at 100 Kya, and everybody else remained clustered together at around 40 Kya. The new dates for the races could not have been further from expectation; it had long been accepted within polite society that the Caucasian peoples were the oldest, being the most technologically advanced and that they had been the colonisers of planet Earth.

The genetic profiling conducted among Australian Aboriginal revealed that they had an unexpectedly elevated level of genetic variance. The levels of evolutionary mutations showed very clearly that these population groups were much older than Caucasians, Mongoloids and Negroids. These surprising findings were the result of detailed examinations of just a small number of mtDNA samples, all taken from pure lineage Aboriginal Australasian people. An increased study size might well have led to a deeper inferred age for the population. The Aboriginal Australian race was not only much older than imagined, but they were pushing back the first emergence of *Homo sapiens*.

Professor Wilson discussed the extraordinary work conducted by Cann, with two scientists then working on a book about the evolution of the human brain, John Gribbin (PhD Astrophysics) and Jeremy Cherfas (PhD Animal behaviour). He explained that the radical information emerging from the project would likely fall foul of the institutional racism existing in the field of anthropology as well as the lack of objectivity than a standard possession of those researching in that area of study.

Though aware that the results would be incredibly controversial, Wilson was willing to offer an interpretation that might at least explain them for the benefit of his two scientific colleagues. From the perspective of Professor Wilson, the first of two viable explanations for the results was that modern humans had migrated from Africa into South East Asia, there encountering other modern humans. The second group of people would have evolved in South East Asia from another population of *Homo erectus*. Interbreeding with them then lead to a hybrid race; the mtDNA discrepancies could then be explained by the existence of two separate lineages both with different dates of origination. This explanation had some additional incidental support from an independent study, headed by James Neel, an eminent US geneticist, that had discovered unprecedented levels of rare *private polymorphisms* (unique genetic variations) among Australian Aboriginals and New Guineans. Surprisingly, Neel had found that while the genetic mixing pot of London offered a private polymorphism rate of about one in one thousand, the supposedly isolated and homogenous

Australasian populations had ten times more rare variants. One would indeed expect to see high levels of these rare mutations if one human lineage had absorbed another. Remnants of polymorphisms from one population would have been forced into the private polymorphisms category as the smaller group was assimilated.

The second suggestion Wilson offered for Cann's findings was perhaps the more radical, even though suggesting Australian Aboriginals were hybrid offspring of humans, and an unknown relative was extreme enough already. The alternate solution Wilson offered to the problem was, "... that *Homo sapiens* originated in Australia, and the spread occurred in the opposite direction."

It was entirely feasible that the results were very much as they appeared to be, representing the emergence of *Homo sapiens* 400 Kya from a small number of *Homo erectus* individuals that had made their way into Australia in the remote past. There was also secondary support for this possibility that had emerged out of the study of DNA base pairs associated with a peculiar ancient virus common to all primates. The sequence of base pairs in humans, instead of being the same as those found in African primates were a match for the sequences observed in Asian primates, leading the lead researcher, Raoul Benveniste, to conclude that, "most of man's evolution had occurred outside of Africa".

In what was further prescience on his part, Wilson also explained to his two fellow scientists that he would expect the interbreeding situation elsewhere to be like that suspected in Australasia, with modern humans absorbing other human lineages, such as Neanderthals. Today we know that this is indeed absolutely the case.

Professor Wilson was all too aware that presenting such controversial research findings from to the scientific community would likely be career suicide. There would be claims that the physical evidence was just not there to support the genetic findings. Truthfully the lack of human fossil finds in Australia is explainable by an absolute lack of effort from scientists to seek out viable dig sites, and further that Australia is not a great continent for sedimentary rocks and volcanic layers, making it incredibly hard to acquire well-dated finds. Then there is the question of what evidence to look for as there is no guarantee that the earliest humans on the continent were using stone tools. East Africa offered much more potential for quickly made discoveries and readily datable artefacts.

At the time that Wilson gave the statements featured in Gribbin and Cherfas book, *The Monkey Puzzle,* it was the very start of the 1980's, and nobody in the academic community suspected anything other than *Homo sapiens* emerging from among *Homo erectus* populations in East Africa, between 100 – 200 Kya. We can imagine how controversial it would have been to suggest that *Homo erectus* populations crossed the Wallace line into Australasia hundreds of thousands of years ago, there evolving into modern humans. Even offering the conclusion that *Homo sapiens* diverged from the common ancestor shared with other hominins early enough to be modern humans by 400,000 years ago, this alone would have caused an uproar. Long before any mention of another archaic hominin revealed in this revised human ancestry, there would likely not have been any scientists still listening. We will never know for sure what might have been. The possibility of an Australian origination for humanity was soon dropped, despite two papers being published based on the study of the 112 samples, *Length Mutations in Human Mitochondrial DNA* (1983) and Polymorphic Sites and the *Mechanism of Evolution in Human Mitochondrial DNA* (1984), no mention of the anomalies was made.

Cann and Wilson moved swiftly on with the Out of Africa Theory, the divergent results were shaved off, and when they published *Mitochondrial DNA and Human Evolution* in 1987, the mtDNA lines had all been shifted towards an African origin. Perhaps strangest of all, among the closing statements in the 1987 paper is the following remark:

"If there was hybridization between the resident archaic forms in Asia and anatomically modern forms emerging from Africa, we should expect to find extremely divergent types of mtDNA in present day Asians, more divergent than any mtDNA found in Africa. There is no evidence for these types of mtDNA among the Asians studied. Africa is a likely source of the human mitochondrial gene pool."

We will recall that Wilson had specifically highlighted anomalous divergence among Asian and Australian mtDNA samples, at levels well beyond the African samples. How could there now be no evidence of previously claimed discoveries?

It might be tempting to think that the anomalies had simply been reconciled by further analysis, but this seems unlikely as the unexpected diversity among Asians, relating to the private polymorphisms, also appeared in a later paper by Ranajit Chakraborty (1990), emerging from his study of mtDNA sampled from six disperse Asian populations. More importantly a genetic study that was carried out by two geneticists in Geneva, Laurent Excoffier and Andre Langaney, specifically to challenge Cann and Wilson's theory (and to place Caucasians back on top) also found themselves in the strange position of having to place an Australian Aboriginal lineage in the centre of their gene map.

"Type 69 (an Australian lineage postulated to be ancestral) is the central type of the 133 types, according to our definition, and thus presents all restriction sites in their more frequent state. It is interesting to note that our partial phylogeny was clearly arrayed around type 69, to which types found in various continents connected directly. Even if this phylogeny would suggest to some people that the mitochondrial Eve was Australian, we would rather consider that these potentially old types were present in continental groups before the different population splits."

In 1991, Wilson, the senior researcher, passed away just before the publishing of the academic paper that he is remembered for today. Sadly, we shall never know what he would have said about the many revelations since then that support his ponderings on Australia as a cradle of humanity.

We have already seen that everything these two scientists

deduced from a genetic research project over thirty years ago is today supported by findings made at new archaeological sites and from additional genetic research projects. I am sure that had Cann and Wilson carried out more extensive sampling in Australasia they would eventually have reached the actual divergence date of *Homo sapiens*, now known to lie somewhere between 700 – 900 Kya. It's hard not to feel perplexed by the fact the research community has still not taken Professor Wilson's initial speculative conclusions more seriously in light of so much corroborative data. Surely there is now every reason to suspect that Australasia represents the best candidate continent for the original home of *Homo sapiens*, with indigenous Australasians being ancestral to all modern humans?

My question above is, of course, rhetorical; we have already seen the compelling evidence and come to understand that the bias in the research community leads to problematic data being massaged to fit the OoAT or even total avoidance. As a rule, archaeological evidence that places modern humans in Australasia significantly earlier than 50 Kya is considered to be too controversial and gets ignored. The limited research that is carried out on the oldest discoveries results in reports buried in publications that do not reach the public at large.

There is evidence from across Australia showing that ancient human beings were deliberately lighting fires to clear land (fire-stick farming) before they were supposed to have left Africa. Researchers from the University of Tasmania examined the incidence of fires across the continent, spanning a 350,000-year extended period, with surprising results. The University of Tasmania study found that levels of charcoal dramatically increased alongside levels of eucalyptus pollen around 120 Kya, especially in the Lake George region of New South Wales. The team suggested that this reflected an earlier entry into Australia (than usually understood) by early human populations.

It transpired that the evidence for ancient fire-stick farming near Lake George was not to be a solitary discovery. A core sampling project in North Queensland, headed by leading climate scientist Dr Peter Kershaw as well as noted geologist Jim Bowler, had stumbled upon more signs of humans using fire-stick farming methods incredibly early on in time. Core samples extracted from the Great Barrier Reef covering the last 1.5 million years revealed compelling evidence of a sustained and deliberate burning of rain forests around 140 Kya. The team reached the inevitable conclusion that humans must have entered Australia at least 80,000 years earlier than believed. This finding should have made headlines across the globe, of course, though it never did so.

It is not just evidence of ancient fire-stick farming that has placed humans in Australia earlier than previously suspected. The very same geologist from the Great Barrier Reef core study, Jim Bowler, recognised as an ardent OoAT supporter, has found himself in the uncomfortable position of unearthing further controversial evidence. His detailed examination of an ancient hearth at a site in Point Richie, Victoria produced discarded shells and charcoal that dated the site to around 80 Kya, but certainly no less than 70 Kya.

The common themes between the Australian archaeological sites already highlighted are that they each provide compelling evidence for human activity in Australia earlier than is allowed in the OoAT and that they were promptly dropped soon after being acknowledged as important. The fires that once burned at these sites have long since been out, but it appears that their remnants remain too hot for any scientist to touch.

Archaeologists in Australia have uncovered very few ancient human fossils. There are relatively small numbers of researchers covering an immense landmass, much of which is dominated by inhospitable deserts. Aboriginal Australians almost certainly know of locations where human fossils might be found, but are usually against the desecration of their ancestors' graves. It is for this reason, and others, that we must rely principally on genetic studies to understand the story of ancient hominins in the Australasian region. We do at least have two fossils that will help our enquiries. Incredibly, Jim Bowler again plays a significant role in this case as it was he who discovered the remnants of a fossil skeleton at Lake Mungo, New South Wales, in 1969. The find, a cremated female, was designated as LM1 but has since become known as Mungo Lady. In 1974 Bowler uncovered another set of remains at Lake Mungo; these were those of a carefully buried male designated LM3, but are commonly known as Mungo Man.

Lake Mungo had once been part of a system of ancient rivers and lakes that played an essential role in sustaining life in the interior of the country. Ancient humans would have indeed migrated towards these vital sources of fresh water. As conditions changed hundreds of thousands of years ago, this system began to fail, and arid conditions gradually took over. By around 40 Kya this system of waterways was no longer present; assumedly, humans would have migrated away at that stage.

The burial of Mungo Lady had apparently been an elaborate affair; the remains had been cremated and then pulverised, followed by a second cremation. The ashes were interred in the ground along with a covering of red ochre (a red coloured mineral), sourced from locations several hundred kilometres away. Initial dating suggested the fossil fragments came from a period 25 – 20 Kya. The extreme state of the remains did not allow for any analysis beyond dating and attributing a sex.

Unlike Mungo Lady, Mungo Man's fossilised bones were in a reasonable state of preservation; there had been no cremation or pulverisation. This ancient man had been buried flat on his back with his arms folded; the fingers interlocked at his groin. During the burial process, the body had been sprinkled with a liberal coating of red ochre. Several methods of dating were carried out, and the results gave possible ages falling in a wide range spanning 86 – 28 Kya.

The key players in the story of Mungo Man are Jim Bowler and Alan Thorne. Bowler, the discoverer of the fossil, led those that firmly believed a younger date was appropriate for the remains, perhaps 30,000 years old, but with an absolute earliest date of 40,000 years. Alan Thorne and his associates, responsible for carefully piecing the remains back together and running additional testing, were adamant that a dating close to 60 Kya was most appropriate. It is important to understand that at the time of this debate Aboriginal Australians were claimed to have entered the country just 20 Kya. The scientists on both sides of the Mungo Man discussion were suggesting dates that would have been implicitly controversial for mainstream archaeology.

The debate raged on for decades. Only recently have examinations resulted in a better-accepted date of approximately 43,000 years before present, attributed not only to LM3 but also LM1. Additional physical evidence from the Lake Mungo archaeological site shows us that human habitation of the area certainly dates back at least 50,000 years. This early inhabitation of Lake Mungo is problematic to understand; the area is some 2500km from any potential entry point into Australia, as well as being arid and relatively barren. There remains the question as to why any humans would land on the northern Australian coastline and immediately run thousands of kilometres from the idyllic beaches into the harsh outback to scratch out an existence in an inherently hostile environment.

Alan Thorne and his team had not only reconstructed the skeleton of LM3, but they also managed to do the first ever successful extraction of genetic material from such ancient human fossils. The results of genetic testing of the Mungo remains only evoked more hostility from the scientific community, as it suggested that Mungo Man had a much higher than expected level of genetic divergence. In fact, the genetic separation exhibited in these remains was so great that it was not possible to directly associate Mungo Man with either Africans or Aboriginal Australians. It appeared that the people of Lake Mungo were an unknown human race that had become extinct in a mysterious epoch of pre-history.

One key assessment of the findings from the Mungo Man study appeared in a paper by Gregory J. Adcock, et al. in 2001. The interpretation they gave was that the mtDNA belonged to a lineage now only evident in modern human populations as a segment inserted into one chromosome, designated as chromosome 11 of the nuclear genome. Mungo Man's genetic line was calculated to have diverged away from the most recent common ancestor of all modern human populations. This led to the conclusion that the oldest modern human lineage, based on mtDNA research, was Australian. The examination of genetic mutations in Mungo Man's complete genome suggested that, based on the so-called *molecular clocks* used for dating hominin divergence, he should be around 170,000 years old.

It will come as no surprise, perhaps, to learn that the scientific community rejected all the controversial findings and wrote off the studies as being flawed. The general view was that the results must be from contamination because modern humans only arrived in Australia 50 Kya, and no other hominins were accepted to have entered the country (in the consensus view). The suggestion that any fossils might display the genetic signature of a more archaic *Homo sapiens* population in Australia, perhaps arising from a separate divergence event to other modern humans, well, that was tantamount to scientific heresy.

In a recent set of new tests involving genetic material extracted from Mungo Man, experiments headed up by Professor Lambert of Griffith University found that at least some of the samples from the original studies were contaminated. It appeared that handling of these specific samples had allowed traces of the researchers' own genetic material to get mixed in. Professor Simon Easteal, a co-author of the original Adcock paper, readily accepts that Professor Lambert and colleagues have found contamination, but he feels this does not rule out his team's earlier findings. Professor Easteal suggests that "there may still be ancient DNA in samples of Mungo Man not studied by the researchers", and as such, there is still room for further developments in this case.

The peculiar thing about the findings from the initial examination of Mungo Man's genome is that it produced a conclusion that agrees with the results of more recent scientific studies. We have already seen evidence that other archaic hominins were indeed living in Australia alongside fully modern humans. We have also seen evidence of populations going back to dates equalling that calculated for Mungo Man's divergence. The Adcock paper appears to be at the very least highly prescient, whether it included results from contamination errors or not.

Another find made in the first Thorne-led studies has remained valid and uncontested. The anatomy of Mungo Man is that of a tall and gracile person, rather reminiscent of the northern European people of today. The bone of the skull is 2mm thick. This is unusually thin and frail, very unlike other more typically robust skulls from this early period. Measurements of the bones of the limbs suggested a height of 196cm (6ft 5in), making the deceased much taller than most Aboriginal Australian men, past or present. The average height among Aboriginal Australian men today ranges between 165cm and 172cm, depending on which geographic region is surveyed. This unusual anatomy comes to the fore only when we contrast Mungo Man against another more recent population living 200km south of Lake Mungo.

The Kow Swamp burial ground located in northern Victoria sits close to yet another depleted ancient lake system. Excavations of graves revealed forty sets of skeletal remains dated to a period between 15 – 20 Kya. Despite the relatively early dates attributed to these remains, they are unlike any others of their era. Almost all the Kow Swamp people display morphological features closely resembling those of *Homo erectus*. The skulls are unusually large, notably elongated, and up to 13 mm thick. That is 11mm maore than LM1 and almost twice the average for modern humans. The faces would have been big, broad and projecting, displaying prominent brow ridges and receding foreheads. Perhaps most shocking were the teeth which were enormous and set in large jaws. These teeth were larger even than those of *H. erectus* living on Java 1 million years earlier.

Homo sapiens displaying such archaic features is shocking enough, but considering the fact that fully anatomically modern humans lived just a short distance away at Lake Mungo 20,000 (or more) years earlier makes it even more stunning. Had the Kow Swamp remains been found on an Indonesian island and produced earlier dates they would have been considered a variant of Asian *Homo erectus*. In fact, these strongly resemble the so-called *Homo erectus soloensis* of Java that we encountered previously. They also lived 20 Kya and exhibited transitional morphology reflective of both *H. sapiens* and *H. erectus*. Surely, we can't imagine that among one small group of incoming migrants there would be sufficient genetic diversity to produce multiple morphologically distinct human forms?

Kow Swamp appears to offer further compelling evidence for an evolution of *Homo sapiens* directly from a population of *Homo erectus* living in Australasia. In a profound sense, the Kow Swamp people challenge the very definition of what it is to be a fully anatomically modern human being. The only skull that can be considered more out of place is that of Willandra Lakes Hominid 50 (WLH 50) found not far from Lake Mungo and dated at 30 – 25 Kya. This skull has a cranial capacity of 1450cc, large brow ridges, and is 17mm thick.

Consider here that even in East Africa, the supposed cradle of humanity, purportedly the home of the greatest human genetic diversity ever, there is no evidence of such incredible morphological variation between neighbouring communities. It is also only in Australasia that we find the genetic traces of four human species alongside evidence of several morphologically and genetically distinct *Homo sapiens* populations. We can only imagine that, were Professor Wilson alive today, he might feel emboldened enough to present the history of *Homo sapiens* inclusive of an early evolution of the species close to 700 Kya, arising out from a population of *Homo erectus* that reached Australasia close to one million years ago.

Chapter 7 – Tracing the *Homo sapiens* Migration into Asia with mtDNA and Y-Chromosomes

Per the information provided by the official website of the Australian Museum, archaeological evidence supports modern humans as having reached Southeast Asia by 70,000 years ago. It has taken a very long time and numerous problematic discoveries for this early arrival date to be accepted by that institution. If you read further into the text of the Australian Museum's description of the populating of Asia, you will learn of two competing theories. On the one hand, there is the most widely accepted theory that modern humans migrated from Africa and rapidly colonised the Asian region, replacing any hominins already present. On the other hand, there is the more controversial theory that Asians emerged in the region directly from an existing population of *Homo erectus*.

In 2011, Eske Willerslev of the University of Copenhagen headed up a project which analysed a century-old lock of hair taken from a full-blooded Aboriginal Australian man. The results of this study revealed that, despite popular acceptance of the 50,000 year-long isolation of Australasians, the divergence between Aboriginal Australians and any of their neighbours happened much earlier. This hair revealed that the minimum date for Aboriginal Australians separating from other humans was between 75 – 64 Kya. Further critical reviews of these findings have tended to support the date of 75 Kya, suggesting that inhabitation extended for 25,000 years longer than commonly accepted.

Associate Professor, Darren Curnoe, leader of the Human Evolutionary Biology Lab at the University of New South Wales, voiced his opinions on how the examination of the Aboriginal hair sample had revolutionised understandings. Curnoe explained that the result "powerfully confirms that Aboriginal Australians are one of the oldest living populations in the world, certainly the oldest outside of Africa", also conceding that it suggested humans had been on Australasian land for 70,000 years. Unsurprisingly, Professor Curnoe does his best to place this update into an Out of Africa context. The Out of Africa Theory would now seem to require a group of Africans running wildly through the Levant, Middle East, Central Asia, East Asia, South East Asia and then hurriedly building the world's first boat so that they could get to Australasia as fast as humanly possible. Clearly, this mad dash theory is nonsensical; it goes against everything we know about hunter-gatherers and their slow migrations as they seek out new resources.

My cynicism aside, the important fact we must take from this is that all populations in Asia are younger than Australasians. It's hard to see how an older population can arise from a younger population. There is also no evidence of separate waves being responsible for colonisation across Australasia and Asia, which might have explained the discrepancy. The theory that a small splinter group populated Australasia after breaking away from a larger population living in Asia now becomes hard to take seriously. Logic would suggest that we need to flip that equation around and start looking for evidence suggestive of migration out of Australasia between 75 – 70 Kya.

We are unlikely to find a fossilised ship – that would certainly be the best evidence for a long forgotten migration by sea. Yet, we do find the oldest representation of a boat in rock art at Kimberley (Northern Australia), and it is suspected to date to between 17 – 50 Kya. With limited physical evidence to go on, we must again turn to the genetic data. We must delve into the study of Haplogroups to complete our understanding of human origins and migrations. This is a rather complex area of evolutionary biology, one that I am not qualified in, having only recently familiarised myself with the subject. I apologise in advance if my interpretations are at all difficult to follow.

We should start by gaining an understanding of the key terms that will be used here. A *haplotype* is a group of genes inherited by an organism from a single parent. A *haplogroup* is a collection of similar haplotypes that share a common mutation in their ancestry. More specifically, a haplogroup is a combination of alleles at different chromosome regions that are closely linked, and that tend to be inherited together. A haplogroup pertains to a single line of descent, usually dating back thousands of years. Membership of a haplogroup, by any individual, relies on a relatively small proportion of the genetic material possessed by that person. We will often be using the *Hg* abbreviation when referring to haplogroups.

When you look at how haplogroups cluster in geographical regions you can understand why they are part of understanding human origins and our migrations; shared haplogroups are a feature of closely-related human groups. All African mtDNA arises entirely from seven sub-lineages of haplogroup L, and HgL is itself a vanished line associated with the last common ancestor of all humans on Earth. HgL is calculated to have emerged between 233 – 151 Kya. The accepted oldest daughter lineage is L0 which is closely associated with the KhoiSan sub-Saharan Africans and given an age of up to 190,000 years, while outside of Africa mtDNA arises almost entirely from the two ancient lineages HgM and HgN that diverged out of the HgL3 line.

Evolutionary biologists admit that it is unclear where L3 first appeared. The timing of its formative mutation is calculated to be between 104 - 70 Kya. Evolutionary scientists suspect that HgL3 emerged in East Africa but did not spread from the continent until between 74 - 70 Kya. The age of both the HgM and HgN lineages, at around 70,000 years, seems to suggest that HgL3 mutated twice, immediately after the exit from Africa. This immediate double mutation theory is hard to swallow, at best. Even one significant change right at the moment of the exit is an amazing coincidence. There is an equal possibility that the HgL3 mtDNA lineage was itself brought to Africa during an ingression by migrants coming from an older population in Australasia, some carrying HgM or HgN while others carried HgL3. It is this intriguing possibility of an HgL3 introgression event in Africa that we will be exploring further.

Palaeoclimatologists carefully modelled the climate of the Earth at the time associated with the OoAT migration, 74 Kya. This revealed that severe climatic problems were in effect that would have made movement through North Africa, the Levant or the Middle East almost impossible to imagine. Between 74 - 70 Kya Northern Africa – indeed, the entire Northern hemisphere of Earth – was suffering a virtual nuclear winter. Paleoclimate modelling reveals that from 70 – 60 Kya the era claimed as the timeframe for modern human dispersals out of Africa incorporated the most extended drought period in the last 125,000 years throughout northern Africa, Saudi Arabia and the eastern Mediterranean. Professor Timmermann, a key researcher in the climate modelling project, made a very poignant remark on the effects the drought would have had on humans: "Walking into the Arabian Peninsula around 70,000 to 60,000 years ago, would have been a bad choice."

The ancient climate data conflicts with claims that HgL3 daughter lineages, HgM and HgN, began to expand geographically from Africa around 74 - 70 Kya as climate started to improve. We will explore in more detail the cause of the severe climatic events undergone at that time and the limitations they put upon migrations, but first, we must explore more of the available haplogroup data.

One cause of confusion for academics is the presence of ancient lines of the HgM and HgN haplogroups among the indigenous peoples of Australasia – about as far away from Africa as you can get. Their presence in Australia led many evolutionary biologists to conclude that both of these lineages must have emerged somewhere in Asia soon after the possible migration out of Africa. It has also been theorised that HgM and HgN arose somewhere between East Africa and the Persian Gulf, in the paper Torroni et al. (2006). One example of a major hole in these theories on where the mutations occurred is that in Australasia we find mtDNA lineages that mutated from haplogroup HgN which are unique to this region: HgO and HgS. Why did the supposed Asian ancestors of the Australasians lack these haplogroup lineages?

It is not only the matrilineal mtDNA chromosomes that call into question current population migration theories. There is also relevant information to be considered regarding the patrilineal (male reproductive line) Y-chromosomal haplogroups with their many subclades, especially the HgCT subclades. In Australasia, the most common male lineage haplogroups are variants of HgC (or C-M130). Geneticists consider haplogroup HgC to be extremely ancient, arising directly from HgCT, one of the three founding lineages with haplogroups HgA and HgB.

Both HgA and HgB are unique to African populations, and HgC is found only outside of African populations. Put another way; all Africans stem from the ancient haplogroups A and B while all non-Africans arose from the equally old HgCT lineage, a line that lacks a clear link to African populations. Evolutionary biologists usually suggest that HgC must be a slightly newer line that arose from a mutation which occurred immediately after the first African migrants left their continent. Yes, incredibly, just as we were told to believe with mtDNA, the Y-chromosomal haplogroups of Africans supposedly mutated the moment migrants began their sprint to Australasia. It is never explained to us why contemporaneous mutations occur on both the male and female lines at the very same time, despite the difference in time it takes for mutations to occur. It is also not clarified why mutations like these occur at the moment African people step over an imaginary line separating them from the rest of the Earth.

Unlike the African chromosomal lineages that were supposed to be foundational to all individuals on Earth, but are notable by their complete absence beyond Africa, haplogroup C has a surprisingly high geographic spread. We also find that in Australasia, where the people are considered to be closest to the hypothetical African root population of 70 Kya, there are unique mutational subtypes of the primitive C-M130 haplotype. The uniquely Australian lineages, designated as HgC-M347, HgC-HgM210 and HgC-M347 (previously these were named HgC4, HgC4a and HgC4b), also happen to be the most common haplotypes across that continent, represented within the genome of 65.74% of Australian Aboriginal men. Before the European colonisation men in Australia, almost all carried one of two main variants of this haplogroup, rather suggestive of two distinct lines of ancestry. Another variant, HgC1b2a (previously C2), is found among Melanesians but not found in Australia, so can be assumed to represent a mutation which occurred after a separation caused by rising sea levels. A smaller percentage of men, around 3%, still carry the oldest variant HgC-M130. It seems incredible that after just a short period on the evolutionary timescale a tiny group of African migrants had entirely moved away from their genetic origins and fallen into two or three clusters of uniquely Australasian haplotypes. The observed clustering strongly suggests there were genetically distinct founder populations.

Per the *American Journal of Physiology*, several more haplogroups are specific to Original Australasians, including HgC-M347, HgK-M526* and HgS-P308, all of which are believed to be well over 40,000 years old. It seems incredible that there is no sign of any of these haplogroups in Asia. How can we be asked to believe that early migrants into Asia founded the Australasian populations? Are we to believe these many local mutations arose in a relatively short time, despite the lengthy mutation rate required for changes to the Y-chromosome?

In contrast to mitochondrial DNA, which has a short sequence of 16,000 base pairs and mutates frequently, the Y-chromosome is significantly longer at 60 million base pairs and has a lower mutation rate. The mutation rates of mitochondrial DNA are ten times higher than observed in nuclear DNA (this includes the Y-chromosomal information). Mutations occurring on specific marker sites of the Y-chromosome are estimated to occur once every 500 generations per marker. How long would it take for all these Australasian mutations to occur? How much longer for these mutations to become the most common haplogroups on the entire continent?

As worthy as my above questions might be, a much more important issue arises once we learn that HgC lineages are common across most of Asia. Why is haplogroup C so common among these younger (than Australasian) Asian populations? Undoubtedly, daughter HgC lineages must have emerged from an older founder population, situated nearby, that carried the precursor variant.

The commonality of the HgC lineage varies hugely across the Asian region, and beyond; it is prevalent in various geographical areas. HgC is at its highest levels in the following areas and populations:

Polynesia (40.52%); Heilongjiang of northeastern China (Manchu, 44.00%); Inner Mongolia (Mongolian, 52.17%; Oroqen, 61.29%); Xinjiang of northwestern China (Hazak, 75.47%); Outer Mongolia (52.80%); and northeastern Siberia (37.41%).

HgC is also present at a lower commonality in other regions, extending longitudinally from Sardinia in Southern Europe all the way to Northern Colombia, and latitudinally from Yakutia of Northern Siberia and Alaska of Northern America to India, Indonesia and Polynesia. This spread certainly fits a model in which *Homo sapiens* are living in Australasia first, before eventually emerging to populate Asia and then the world beyond.

Rather than applying convoluted thinking that might allow us to fit the haplogroup data into the usual OoAT, let us first consider a straightforward and eminently logical scenario. Let us imagine a group of Australasians carrying HgCT and daughter lineages. Some of these people – not all of them – migrate west across the Wallace Line into South East Asia around 74 Kya. The migrants entering the South East Asian region would be affected by the environment, and this would result in accelerated selection and mutation, especially if they interbred with any hominin populations present in that area. Just as we would expect in this scenario, South East Asia is found to have the most variants of HgC. As we move out of the region, the trace begins to fade, and then new and more recent genetic lineages become increasingly widespread.

The evidence that HgCT emerged from Australasia is beyond compelling. We already know that Australasia has recently been accepted by academics to hold the oldest human populations outside of Africa. One article in the journal *Nature* on the global distribution of HgC offered the opinion that HgC provides important clues about the early colonisation of Asia by anatomically modern humans. I couldn't agree more with their thinking. The same scientists involved in HgC research also explored evidence in the fossil records of East Asia, highlighting dental traits that suggest East Asians descended from Southeast Asians migrating across the Sunda shelf. We should also recall the fact that modern East Asians share shovel-shaped incisors with the archaic hominins of Indonesia, providing additional support for evolutionary links between the two.

One of the most important discoveries in recent years, the Ust'-Ishim femur, also fits the evolutionary scenario discussed above. Discovered in a Siberian riverbank approximately 1200km to the north-west of the Denisovan Altai cave, this femur transpired to be that of a 45,000-year-old man from an unknown modern human population wandering around in the absolute middle of nowhere, distant from any known archaeological sites. Testing by the Max Planck Institute revealed the oldest complete genome yet recovered from such ancient human remains. Not only was it deduced that the individual was part of a group that left no surviving lineage today, but this person had haplogroup K-M526, one of the haplogroups unique to Original Australians (carried by 27% of Original Australian men).

There is no doubt that the mysterious population in Siberia, of which Ust'-Ishim was a member, had travelled up from South East Asia. The scholarly understanding offered is that this migration began around 55,000 years ago (the early stages of a 10,000-year warm period). This individual carried the 2% Neanderthal DNA normal for modern Eurasians. The Neanderthal DNA in the Ust'-Ishim genetic profile entered his genome between 58 -52 Kya, so this places the interbreeding event at the start of the journey, in South East Asia. This finding runs contrary to those research scientists who have argued that modern humans interbred with Neanderthals in the Levant or the Middle East immediately after leaving Africa around 70 Kya. The levant interbreeding hypothesis has itself suffered a fairly major blow, with excavations of a Neanderthal site in western Iran strongly supporting a modern human replacement of Neanderthals in this part of the world around 50 Kya, only after the end of the intense regional drought and well after the supposed emergence and replacement events associated with OoAT (remember here that the oldest modern human skull in the Levant is 55 Kya).

You might very well wonder why *Homo sapiens* would suddenly emerge from Australasia and migrate into the many lands beyond. We know some small groups of modern humans were already out there – teeth in Chinese caves tell us that – but we seem to see more archaic human forms dominating Europe and Asia up until that time. If people were in Australasia for such a lengthy period, hundreds of thousands of years, why do we not see significant numbers of *Homo sapiens* migrating out of Australasia to colonise the planet until after 75,000 years ago?

It is apparent that *H. sapiens* were moving around East Asia and South-East Asia before 75 Kya, as were their cousins the Neanderthals and Denisovans (and others). We have already seen fossil remains across this region that pre-date the era of global colonisation. If we consider the levels of Denisovan mtDNA among Papuan populations today, we would expect to see Denisovans and their close relatives stretched from Northern Sahul across into most of Sunda. We also know that novel hominin species lived on islands between the two continental plates on both sides of the Wallace Line, including *Homo floresiensis* and *Homo erectus soloensis*.

Consider that 74,000 years ago global temperatures were around 6 degrees cooler, on average, than today. During the colder periods of the ice-age hominins would have highly favoured land situated in the warmer equatorial belt. South East Asia and Northern Australian would have been prime real estate. We know that Neanderthals were able to tolerate cold environments better than modern humans, allowing them to colonise regions well beyond South East Asia. Despite the hardiness of Neanderthals, it is likely that they would also have held lands within the warm equatorial belt; survival in temperate climes with plentiful food is a natural draw for all humans.

The *Homo sapiens'* expansions into Asia occurring after 74 Kya were, then, new waves of migrants. They represented a repopulation of territory, not a first colonisation. To understand the position I am arguing, we need to explore the evidence of a terrible catastrophe, an event that rocked our world. Right in the middle of the Sunda landmass on what is today Sumatra, 73,880 years-ago (scientists claim a dating uncertainty of 640 years and 95% confidence), the Lake Toba super-volcano roared into life. This huge Indonesian volcano produced the most powerful eruption of the last 2.5 million years. This disaster was beyond all imagining; seven trillion tonnes of volcanic material were ejected during the event which lasted for 14 days. Some 800 km^3 of ash rolled across South and Southeast Asia as well as the Indian Ocean. Millions of square kilometres of land became covered in debris.

Within 150km of the caldera, the 'pyroclastic zone', there would have been a scorched and barren hellscape, with flows of searing hot gas and ash flowing outwards faster than the speed of sound. Nothing in the path survived. Intense lava flows followed the initial blast, with over 2800 km³ of magma produced. The sky locally would have darkened from day to night as the ash became clouds in the stratosphere. Earthquakes beyond any known today would have preceded and accompanied the initial eruption and the eruptions which followed; many would have been in high 9s of the *Richter Scale*. We can imagine the associated tsunami waves, far stronger than those of the 2004 Asian tsunami, rolling across the oceans and laying waste to coasts in all directions.

Being a few hundred kilometres from the blast would only have meant a slower death, but death all the same. Not only ash rocketed skyward during the explosion, but also vast quantities of sulphurous gases. When the next rains fell, they were black, powerfully acidic and toxic to all plant and animal life. For thousands of kilometres' thick blankets of ash rained down in immense amounts. Evidence from core samples suggests that over 3 metres of this material covered parts of distant India. The prevailing winds were to the north-west but, thousands of kilometres to the east, ash even rained into the South China Seas.

Climate models show us that global temperatures had been in decline even before the Toba eruption. The dust pumped into the atmosphere blocked sunlight and then accelerated and deepened the cooling trend. The greatest effects were in the northern hemisphere where average temperatures fell by 3-5°C. This temperature range may not sound like very much, but consider that temperatures were not altered uniformly across the northern hemisphere. For example,

Greenland experienced a drop of 16°C. We can imagine that while some areas would have had very little impact, there would have been severe and intense temperature drops wherever the wind carried the greater part of the dust. We need to understand the sensitivity of our climate; even a temperature drop of between 1-3°C can provoke a shift from moderate to extreme conditions across usually temperate regions like Central Europe. All cold-sensitive vegetation would have suffered from the temperature changes, but all plant life would have suffered from the reduction of sunlight, calculated to be around 25% (due to the reflective dust particles). Scientists have estimated that as much as 75% of all plant life in the northern hemisphere died off because of the Lake Toba 'doomsday' event. Even algae and phytoplankton would have been devastated by the effects of the explosion as it rolled onwards into the following decades.

The equatorial region of South East Asia would have been the most densely populated region on Earth at the moment of the Lake Toba event. The catastrophic effects and after-effects would have obliterated human (and animal) populations for several thousand miles. The cooling climate was accelerated into an intense cold period that would last for two thousand years without let-up. Hominins had almost been wiped out across Continental Asia. We know today that the Neanderthals, Denisovans, Hobbits and *H. erectus* entered terminal population decline after this catastrophe. Each lineage faded away to extinction over the following 30,000 years. The final interbreeding events recorded in the genetic profile of modern humans suggest that by around 40 Kya there remained no other distinct populations.

With Lake Toba firmly in mind, we should look at inter-breeding in a new light. The remnants of the various hominin populations would have been forced to migrate; the best option would have been moving further south, deeper into Australasia. Groups of survivors on the wrong side of the Lake Toba kill zone would have been forced to move west. This may have been tolerable for Neanderthals as they had the benefit of genetic advantages acquired by past generations of cold-dwelling ancestors, and they had communities already in Europe to join. It may not have been an option for modern humans to walk into the frozen wastes of glaciated Europe; they would have lacked both the genetic advantages and the survival knowledge essential for such a region (they perhaps even had no concept of clothing). We do know that somebody was still moving around in parts of Southern India after the eruption as stone tools have been found both below and above the ash layer at sites such as Jwalapuram. Perhaps some of these artefacts belonged to survivors making their way to the west. We will never know.

The global population of hominins is understood to have collapsed after Toba, with estimates of around 10,000 adult hominins remaining alive on Earth. Communities living in central and southern Australia would have likely absorbed small numbers of survivors moving southwards; this might have included a range of displaced human types including Denisovans, Neanderthals, *Homo erectus soloensis*, *Homo Floresiensis* and perhaps many others. This merging of any of these lineages would have given the small surviving population a significant boost in genetic diversity and access to the total of genetic advantages this human family had gained while separated into distinct isolated populations. The greatest numbers of refugees congregated in the less impacted Denisovan territory of North East Sahul (modern New Guinea), remaining distinct from modern humans until 44Kya.

This volcanic disaster was almost an extinction event for archaic hominins. Instead, it would eventually ensure a unification of all human types into a single human race. We are them, and they are us. The archaic-looking skulls preserved at sites including Kow Swamp and Wilandra Lakes all silently speak to us of this fusion. We had all come from a single lineage, diverged down separate evolutionary roads and, eventually, coming back together as one.

It would take about 20,000 years for humans to recover their numbers after the catastrophe but we now know that, as the climate improved around 60 Kya, a new wave of fully modern human hybrids made their way across the seas from Australasia and into Asia. This time there would be no stopping them. The dating of this event I offer here has emerged from the genetic research of Griffiths University, claimed as the first genomic study of indigenous Australia, in which it was revealed that the ancestors of Asians and Europeans parted ways with those of modern Australasians around 58 Kya (as the strongest dating in a period of divergence that ranged from 72 – 51 Kya). The dating of the colonisation to 60 Kya has been confirmed by the research work of the acclaimed geneticist, Peter Forster. Incredibly Forster also reveals the approximate number of colonists involved was a little less than two hundred, a very small number when you consider that these few were the ancestors of all Asians, Europeans and indigenous Americans.

We have seen powerfully compelling evidence that an Australasian founder population colonised Asia and then Europe. No other explanation can make sense of the genetic data and archaeological finds, especially once climate history is factored into the equation. The evidence we have explored makes sense of the notable genetic split between modern non-African populations and modern Africans, observed on both the matrilineal (mtDNA) and patrilineal (Y-chromosomal) lineages. We can also understand why African and non-African populations evidence different levels of diversity. When Lake Toba exploded, it killed almost every human in the surrounding region. Whatever difference had existed was significantly reduced. There is now just one last puzzle to solve: the story of modern humans in Africa and the role they played in global colonisation.

Chapter 8 - The Human Migration into Africa

As scientists came to understand the magnitude of the Lake Toba eruption they also began to recognise that this ancient volcanic disaster was the most likely cause of the bottleneck recorded in the human genome. The timescales were in accord: 74 Kya humans had suffered a drastic die-off event right at the time of the biggest catastrophe known during the last two million years. It was a coincidence too incredible to ignore, but a level of discord would soon emerge. The Lake Toba revelation eventually came into conflict with central tenets of the OoAT, specifically conflicting with the belief that modern humans were entirely limited to Africa when the volcano erupted.

In 2013 research scientists completed initial research into the effects Toba had on people living in Africa 7000 km west of the volcano. Sediment cores extracted from Lake Malawi and examined by Christine Lane, a geologist at the University of Oxford, showed that there was no local environmental disaster at the time when material from Toba sank to the bottom of the lake. Despite significant ashfall, even at this great distance from the eruption, the effects had been negligible and short-lived. Modern humans in Equatorial and Southern Africa apparently evaded any high impacts from the climate changes and would not have died off in any significant numbers. This finding at Lake Malawi is perhaps not all that surprising to us; we will remember here that most of the devastation caused by Lake Toba was limited to the northern hemisphere of the planet. This realisation that the climatic disasters had spared Africans seemed to spell the end of the Lake Toba/bottleneck correlation, no matter how obviously correct it was.

We can fairly assume that Northern Africa would have been devastated by the ongoing cooling event, and perhaps a significant number of hominins and modern humans there did, in fact, die off, including various unknown archaic hominin forms. Conversely, any communities able to migrate south to the equatorial region before the plants and animals died off would have been quite safe. There is every reason to suspect that the greater number of hominins alive after the Toba eruption (including *Homo sapiens*) were Africans, principally located in South Africa. African humans lost some of their diversity during the climate shift, but they would be expected to show retention of high levels of the former diversity when contrasted with humans in South East Asia (even though the latter group benefitted from hominin interbreeding). Today we observe that the highest levels of genetic diversity exist among South and East Africans on the one hand, and among Australasians on the other. Everything about comparative studies suggests these are two founder populations that survived a bottleneck event in their separate, isolated and environmentally distinct refuges.

Stanley Ambrose, a Professor of Anthropology, explains that when human populations were reduced in size and then isolated in refuge areas due to the Toba eruption, there would have been an initial loss of genetic diversity and preservation of a small random subset. If these founder populations remained small and isolated for many generations, the phenomenon known as *genetic drift* would lead to a randomised loss of alleles and a fixation of others, further reducing genetic diversity and increasing the differences between the populations. There would also be changes due to adaptations to local environments which, in the case of populations in widely different environments (in our case being Africa and Australia), would be a very significant factor.

Because the current OoAT paradigm in evolutionary science does not recognise South East Asia as the homeland of the largest population of modern humans on Earth pre-Toba, scientists are now struggling to make sense of a population bottleneck caused by Lake Toba, when it did not impact the greater part of Africa. The maddening conclusion offered by scientists (of the OoAT faith) is that, rather than the bottleneck relating to the most devastating catastrophe of the last 2.5 million years, it must be "something else that happened at that time." There is not a moment spared even to consider the possibility that their overarching theory might be flawed, even though it does not fit the evidence in front of them. There are none so blind as those who do not want to see.

I imagine by now you are wondering how there came to be modern humans in Africa to survive the Toba catastrophe. After all, hasn't this author gone to great pains to paint a picture in which modern humans evolved in Australasia, and only in Australasia? Yet here I am talking about Africans already present on their lands well over 74,000 years ago. How does this make any sense?

We have continually seen academics refer to modern humans arising in Africa 200,000 years ago, yet, conversely, the data we have explored points towards multiple human forms emerging between 900 – 600 Kya in Australasia. It may have seemed that I was conveniently ignoring Africa and the wealth of archaeological evidence that places modern humans on that continent long before the migration events I have associated with Lake Toba. There is no doubt that *Homo sapiens* were in Africa at least 160,000 years ago, possibly before that. The genetic rift between modern African populations and non-Africans fully supports the conclusion that non-Africans are the children of the Toba disaster. There is every reason to believe the evidence suggesting that most Africans have an ancestry going back to an earlier (than Toba) human population.

Genetic divergence between individuals in a single population (*intragroup variation*) arises much faster than does separation between members of isolated groups (*intergroup variation*). With that in mind, we would expect Africans and Australasians to be very close in their relationship had they been separated by only 70 – 50 thousand years. In fact, we have already noted that they are surprisingly divergent in these respects. We should consider here that the aforementioned 'founder effect' is partly to blame for this. However, it does appear that the split between Australasians and Africans must be due to more than Toba alone. We require more time to explain it thoroughly.

The American journalist and researcher of racial history, Robert Lindsay, gives us a good sense of why I say we need more than 70,000 years and a founder effect if we are to explain the split between Africans and Australasians. In his extensive investigative work Lindsay reaches a conclusion that "If anyone is evolutionarily on their way to becoming a separate species or subspecies, it is the Aborigines and the Papuans of New Guinea. The distance between them and Africans is greater than the distance between any two human groups."

What we need is more precise dating of the split between the ancestors of modern Africans and modern Australasians. Let us turn to one of our earlier research sources, Rebecca Cann, for her assistance in dating this divergence event. After completing her PhD studies, Cann moved her work back behind the consensus line of African origination. Cann's papers on human genetics do not speak of 400,000-year-old *Homo sapiens* or any invasion of *Homo erectus* into Australia; there is not even discussion of hybridization. Cann currently only claims that "after some African samples, it appeared that the most mutationally divergent people were Aboriginal Australians and Papua New Guinea donors". This present opinion on Australasians seems to be a significant turnaround from the 1980's findings, even so, we must remember that she acknowledges that only a few African samples exhibit sufficient divergence to compete with that observed in Australasian specimens. We also know that she has not factored in the effect of Toba on the Australasian samples. Despite Cann's revisions, it is in her research work that we find clues that narrow down the timing of the separation.

During an interview with journalist Alasdair Wilkins, Cann makes it clear that her research team calibrated their molecular clock based on already *knowing the exact amount of time* Australasians had lived in *isolation* far from their *African ancestors*. With this stated assumption bias, showing flawed and incorrect understanding on all three italicised points, Cann's team came up with a molecular clock. This clock assumed that all novel genetic mutations observed in their Australasian samples must have been the result of a small group of Africa-originating *Homo sapiens* living alone for 50,000 years. The reality, of course, is that Australasians were closer to two million years out of Africa, had been on their lands for hundreds of thousands of years, and had not been isolated for any significant part of that time. From this completely confused start, Cann's team concluded that the base of the family tree was African, with a divergence event for *Homo sapiens* away from other hominins around 200,000 years ago.

If Cann had known that hominins crossed the Wallace Line one million years ago and had been aware of the Sima de los Ouesos proof that placed *Homo sapiens* divergence around 800 Kya (between 900 – 700 Kya), I am sure that she would never have suggested anything like her model described above. The biased assumptions assured that the levels of mutation in Australasian samples were irrelevant; any amount was to be considered perfectly normal for a population isolated for 50,000 years. There is a common expression in the scientific community on the matter of flawed research assumptions: *garbage in, garbage out*.

How does Cann's folly help us with our desire for better dating? Consider now that we already know that the real base of the *H. sapiens* family tree is around four times earlier than Cann's team calculated, close to 800 Kya. This suggests to me that Australasians have been separate from their African relatives for four times longer than assumed, which is 200,000 years. This revision to Cann's study produces an incredibly satisfying dating; it is precisely the dating we have seen for the first appearance of the *H. sapiens* race in Africa. This date also matches the physical evidence of the first modern humans on the African continent.

Evolutionary biologists recognise that the oldest population group in Africa is the sub-Saharan hunter-gatherer community once known as Bushmen, now more accurately termed *KhoiSan* (the Khoi and San peoples). Indeed, it is the samples from KhoiSan individuals that Cann was speaking of when she talked about some African samples seeming to have higher diversity than those of Australasians.

In September 2016, a paper appeared in *Nature* authored by David Reich, a geneticist at Harvard Medical School, on his team's examination of genetic evidence for human ancestry. Reich and his colleagues had assembled a database of genomes from all six inhabited continents containing 300 high-quality genomes from 142 populations. Their first finding was that all humans stemmed from one ancestral population, something that we have seen supported throughout our investigation. Genetic research also concluded that the ancestors of the KhoiSan began to split off from other humans around 200Kya, becoming isolated by 100,000 years later. As a side note, the scientists also suggested the likelihood that by 200 Kya people had already developed modern cognition and behaviour, including language.

Reich's findings are nothing short of incredible. They include everything we would have expected to see if our hypothesis were correct. We have evidence that fits the scenario in which one community of Australasians separated 200 Kya before heading to Africa. We have already seen evidence that suggests earlier hominins sometimes moved back and forth between Asia and Africa. At the time of this separation, approximately contemporary with the hypothetical genetic Eve, it is likely that everybody carried the HgL mtDNA lineage.

It is interesting to discover that around 195 Kya the global climate entered an extended period of extreme cold and dry conditions, stretching until around 130 Kya. Genetic studies of modern human DNA tell us that at some point during this grim period of harsh climate human populations plummeted from around 10,000 breeding individuals down to perhaps just 600. It is possible that the onset of this dramatic climate shift played some role in the exodus of the African ancestors from Australasia. Were these people seeking and hoping for a better situation somewhere over the horizon?

Chapter 9 – A Forgotten Exodus, or two?

The genetic profile of modern Africans confirms to us that their ancestors separated before the interbreeding with Neanderthals and Denisovans. They were gone from Asia before Lake Toba erupted. Modern humans in Africa do still possess genes from another archaic human relative. Bohlender's team detected traces of a mysterious Denisovan-like hominin in the genetic profile of his African reference population, the KhoiSan. The scientists discovered that there was an excess of shared derived alleles between KhoiSan, Neanderthals, and Denisovans which they understood to have stemmed from a yet unknown additional human lineage closely related to Neanderthals and Denisovans.

Analysis of the genome of the KhoiSan and sub-Saharan pygmy populations, led by Michael Hammer, had previously suggested the presence of perhaps 2% archaic DNA. The study had concluded that genes from an unknown hominin population, one that split away from our ancestors about 700,000 years ago, were present in the genetic profile. We already know that this dating fits very well with recently revealed findings that have dated the divergence of *Homo sapiens*, Denisovans and Neanderthals from their shared ancestor.

There is a beautiful correspondence between so many of

the studies that we have considered, well beyond the workings of mere chance. Let's squarely face the elephant standing in the room. If the KhoiSan ancestral population interbred with hominins very closely related to Neanderthals and Denisovans, both being groups resident in East Asia and Australasia, then they must have interbred with them before arriving in Africa. There is no evidence that any of these human cousins were resident on the African landmass.

In Australasia, we find clear evidence of at least three archaic hominin populations, all of which interbred with *Homo sapiens*. One of the three groups is a mysterious close relation to Neanderthals and Denisovans, the lineages that diverged from us by 700 Kya. Could it possibly be that this mysterious fourth lineage detected in the genome of Melanesians is the same one identified in the KhoiSan genome? Even if we are talking about yet another human form (human lineage number five), the KhoiSan ancestors must have been on Sahul to gain genes from a close relative of Denisovans.

Dr Mait Metspalu of the Estonian Biocentre led a team of 98 scientists working on 483 genomes at high resolution taken from 148 populations mostly from Europe and Asia, but with a few from Africa and Australia. Dr Metspalu concluded that all people in Papua New Guinea carry traces of DNA from a group of Africans present there around 140 Kya that mysteriously vanished. The discovery of a missing Australasian population closely related to modern Africans is a perfect fit for our hypothesis that the ancestors of today's African people left Australasia earlier than 140 Kya. The date of 140 Kya would represent the moment when no members of their population remained to breed with, giving the impression that they had all suddenly died out. Before moving to another matter, remember that no humans are even supposed to be in Australasia 140,000 years-ago. I suspect that a much bigger Australasian genome sample would bring more details of this story and take us even further back in time.

What of the migration route taken by these ancestral Africans? Is there any evidence of this population as it moved between continents? In fact, yes there is. Crucially, we have fossils of *H. sapiens* in East Asia which seem to match well with a migration event underway by 180,000 years ago, with *Homo sapiens* still present in that region for several tens of thousands of years to follow. The Chinese fossil finds fit well with a migration event in which some individuals settled at sites encountered on the route. We should also remember here that tools on Sulawesi date between 118 - 200 Kya (Professor Stringer speculates some might be far older). These might have been left during a westward migration.

It is also possible that some of the African founder population sailed west from Australia to Africa, either hugging the coastlines of Asia or island-hopping their way across the Indian Ocean during a time of lower sea levels. There may even have been more islands above sea level for them to utilise. It is a fact of history that the island of Madagascar, just off the eastern coast of South Africa, was first colonised by people sailing from the Indonesian islands. Two thousand years ago people made it all the way to Madagascar using simple watercraft, taking advantage of westerly ocean currents. We can't discount a route by sea for our earlier migrants.

The earliest *H. sapiens* in Africa, *H. sapiens idaltu*, were first identified from fossils discovered in Ethiopia, the Omo and Herto skulls. The Omo and Herto skulls are very alike to others of fully modern humans, but they do display certain morphological differences which led to them receiving the designation of being evidence of a sub-species. Whether they should be considered remains associated with sub-species or not is still hotly debated by academics. A description of these African skulls has been provided by researchers from the Bradshaw Foundation for Palaeoanthropology states:

"The morphology of the skulls display archaic features not found in the later Homo sapiens, but are still seen as the direct ancestors of modern Homo sapiens sapiens."

"However, both the adult skulls are huge and robust, and also show resemblances to more primitive African fossils."

Though it is possible that these African *Homo sapiens* are not entirely anatomically modern, Kow Swamp and Willandra Lakes archaeological sites have both provided skulls exhibiting archaic morphology despite being from recent times. There is every reason to suspect that these features are a direct link to the strange-looking human communities of Southern Australia. It also seems to me that if we do not consider Willandra Lakes 50 to be a sub-species, despite being incredibly robust with heavy brow ridges, a 1590cc brain cavity and a skull as thick as a motorbike helmet, then we should accept the Omo and Herto skulls simply as *Homo sapiens*.

If you look at a map of Africa, you will see that Ethiopia is separated from the Arabian Peninsula by a narrow stretch of water known as the Bab al Mandeb straight. Early *Homo sapiens* could easily have crossed the Bab al Mandeb straight during times of low sea levels, even with simple watercraft. Paddling on a log would have been sufficient. It is incredible that the oldest human fossils in Africa were found so close to a well-recognised entranceway into the continent – just look at the vast size of this landmass!

The highest levels of genetic diversity in Africa centre on the same region of East Africa that has produced *Homo sapiens idaltu* fossils. If incoming Australasian *H. sapiens* interbred with African hominins (perhaps *H. rhodesiensis*) upon arrival 200 Kya, we would absolutely expect elevated levels of genetic diversity in East Africa. It is highly logical to think that the Ethiopian fossils represent the first wave of Australasian migrants arriving in Africa, or hybrids resulting from their interbreeding with a local race. These newcomers would have brought with them the HgL mtDNA lineage. In Africa, after interbreeding, HgL gave rise to haplogroups L0, L1, L2, L4, L5, L6. As previously suggested, in Australasia, HgL became HgL3 and then HgM and HgN.

The dating of the Omo skulls is somewhat controversial, initially dated to 130 Kya. Then, almost four decades later, a second investigation resulted in re-dating to 195 Kya. The revised dates followed a return to the site using hand-drawn maps and old photographs. The new dig added additional bone fragments, found three meters above a layer of ash well-dated to 196 Kya. Richard Klein, an anthropology and biology professor at Stanford University, has suggested that the fossils may have sunk into a lower layer during their history, giving them the false appearance of being much older than in reality. Omo 1 and Omo 2 may yet transpire to be closer in age to the Herto skull at 160,000 years old. Seven examples of *Homo sapiens idaltu* excavated at the site of Jebel Irhoud, Morocco, also date to 160 Kya, suggesting a strong association between this time and the first appearance of modern humans on the continent.

Whether *Homo sapiens idaltu* is indeed simply *Homo sapiens* and whether it be 160,000 years old or 195,000 years old, there is no evidence of any *Homo sapiens* in Africa before 200 Kya and no sign of their ancestors. Consider that *Homo sapiens* diverged from a common ancestor between 900 – 700 Kya, but in Africa, we observe a sudden appearance of *Homo sapiens* in East Africa only after 195 Kya, and in South Africa much more recently than that. The key *Homo sapiens idaltu* archaeological sites are all suspiciously near to entrance ways into Africa that are favourable for watercraft.

We also need to consider evidence for the second ingression of modern humans into Africa, around 73 Kya, involving a more culturally and anatomically advanced human group. We should again consider the two-pronged approach for these refugees relocating after the Lake Toba catastrophe, involving watercraft. Evidence usually associated with easterly movement in the OoAT can equally be interpreted as support for a westerly migration, with groups of people hugging the coastlines of Asia until reaching the Arabian Peninsula, and there crossing the Bab al-Mandeb straights.

It is conceivable that the negrito peoples of South East Asia represent remnants of these displaced inhabitants of Sunda, pockets of Toba survivors that found survival zones. The negrito are a mysterious grouping; typically individuals of this designation are very short, dark skinned, and sport frizzy hair typical of most modern Africans. The communities of negrito people are all quite small and spread across a wide geographical area, from the Andaman islands, through Malasia and Thailand, then on islands of the Philipines.

Though Negritos may be few in number today, their morphology and genomes are incredibly diverse (suggestive of a once highly numerous population). Despite resembling Africans, and having some genetic overlap with African pygmy peoples, negrito's carry very ancient mtDNA and Y-chromosomal lineages that strongly link them to the oldest known Asian populations (archaic hominin genetic lineages). These people may even be related to the diminutive humans of Flores. Certainly, these peoples have the small stature we might expect from a hybridization between modern humans and *Homo floresiensis* and some of them carry genes from an unidentified ancestor. It is now understood that Africans split away from negrito people as early as 70 Kya, with these diminutive South East Asians then entering into isolation for tens of millennia. A final note worth mentioning is that there are strong genetic links between

Negritos and New Guineans.

The second group of refugees likely used their watercraft in a more daring fashion, instead of moving along the coast of Southern Asia, they island-hopped their way to the east coast of South Africa, taking advantage of the westerly currents that move from the Sahul and Sunda towards Madagascar. These two parties would have both carried the HgL3 mutation into East Africa, there interbreeding with existing populations and giving rise to mutually increased genetic diversity.

This watercraft centric theory is not suggested without good reason. The oldest finds of fully anatomically modern humans in South Africa are associated with cave sites along the eastern coast including Klasies River Mouth, Border Cave and Blombos Cave, all of which are situated in a single region. Archaeological investigations at these caves suggest *H. sapiens'* occupation began sometime between 125 – 70 Kya. The cave sites of Southern Africa are a very long way from land routes in or out of the continent, but they are where we would expect boats to land if they had sailed from Sunda (perhaps via Madagascar). Klasies River Mouth, Border Cave and Blombos Cave have produced the oldest known archaeological evidence attributed to KhoiSan ancestors, the first Africans of the L3 mtDNA lineage that appeared on the continent some 70 Kya.

Further telling links exist between the sub-Saharan Africans and Australasians. Australasians are unique in lacking blood groups A2 and B, while the KhoiSan are one of only two significant pockets in Africa with a low commonality of blood type B (compared to other African groups). The other small pocket is directly East from their region, on the coast. These blood types are relatively common among Asian peoples and other Africans. The Aboriginal Australians, Asiatic Negritos and the KhoiSan also all share in a fundamental spiritual belief based on a primal creator serpent, identifiable as Rainbow Serpent mythology.

We can now understand the second wave of arrivals into Africa as being direct ancestors of the KhoiSan and the original bearers of their culture. However, due to interbreeding after arrival, they also acquired genes from the first wave of Australasians (a population that had itself acquired genes from archaic African hominins). This interbreeding extended the depth of the KhoiSan genetic link with Africa by an additional 130,000 years. Even today the KhoiSan continue to retain aspects of ancient Aboriginal Australian mythology, spirituality and behaviour as well as apparent morphological similarities. Discussion of the deep links between the two groups goes beyond the focus of this investigation.

In 2002 archaeologists announced the discovery of an engraved piece of ironstone ochre deep inside of Blombos Cave. The crosshatch pattern of lines carved into the Blombos stone is considered to be the best evidence for the first emergence of modern cognition. Today we also have a shell found at Trinil on Java, dated to 540 – 430 Kya. On the outer surface is a virtually identical crosshatch engraving. We can now appreciate the connection between Trinil and Blombos Cave; modern humans indeed evolved first in Australasia, but they also made a forgotten exodus into Africa.

Afterword

"I know that most men, including those at ease with problems of the greatest complexity, can seldom accept even the simplest and most obvious truth if it be such as would oblige them to admit the falsity of conclusions which they have delighted in explaining to colleagues, which they have proudly taught to others, and which they have woven, thread by thread, into the fabric of their lives." - Leo Tolstoy

All evolutionary models based on the recent evolution of *Homo sapiens* in Africa, recent emergence from Africa, and a replacement of archaic hominins by modern African humans, are fundamentally wrong. The Multiregional Continuity Model, championed by Professor Milford Wolpoff, Professor Xinzhi Wu and Professor Alan Thorne, appears accurate in some parts, certainly in the major premises about gene flow, and the suggestion of a single evolving species that interbred across lineages, *Continuity with Hybridization Hypothesis*. The most notable failing of their model is in missing the central importance of Australasia. The Out of Australia Theory, as formulated by my research colleagues Steven and Evan Strong, offers the correct location and approximately the right timeline but does not discuss the more global picture before *Homo sapiens* emerged nor the migrations into Africa. This existing theory was certainly the foundation for my own work and an invauable starting point, it is essentially correct but did not go quite far enough for my liking.

The most accurate description of my own model would be as a theory of *Homo sapiens* evolution in Australasia with an early emergence, followed by high levels of continuity across locations and ongoing hybridization (including full population absorption). Despite the inherent complexity involved in my model, my preferred name would simply be *The Into Africa Theory of Human Evolution.*

The investigations in this book offer a strong case for a single evolving lineage of humans from around four million years ago and onwards. Offshoots evolved away from this species, but such branches were either reabsorbed or reached dead ends. Most often, sub-species were amalgamated during subsequent interbreeding events. *Homo australopithicus* gave way to *Homo erectus* and eventually from this population some new lineages of modern humans emerged. Were all of these lineages simply variants of *Homo sapiens?* They certainly seem to have been willing to interbreed, and we can now be quite confident that a good number of their children were also capable of reproduction. We should rightly think of these lineages as being *Homo sapiens sapiens, Homo sapiens neanderthalensis, Homo sapiens denisova,* and so on. I am doubtful that there was ever a man and a woman alive on Earth at the same moment who were so different that between them they could not produce a healthy child itself capable of reproduction.

It is clear that the increased diversity gained by separation, isolation, environmental adaptation and then subsequent reabsorption played a pivotal role in producing the morphological variations observed within our human lineage.

This complex picture has at times led experts to erroneously identify new, entirely distinct hominin species among a single human population. A great many of our evolutionary advantages came to us from human lineages that we now consider to be extinct, though a part of them clearly lives on in us. They did the hard work, and we reaped the benefits.

Mystery still surrounds the exact cause of the divergence event in Australasia that occurred between 900 – 700 Kya. Why did a population of *Homo erectus* suddenly give rise to multiple large-brained, anatomically advanced *Homo sapiens* lineages? What might have caused this relatively sudden and profound evolutionary leap? We know that some of the most significant adaptations seem to emerge spontaneously at 800 Kya. There is tantalising evidence that suggests to me we are missing part of the history. We now know much more about what happened and where, but little of the why. These are questions for a later edition within this book series.

We can now recognise the traces of the Australasian colonisation of Africa, Asia and Europe. Australoid appearance gave way to Capoid, Negroid, Mongoloid and Caucasoid. It took time and some devastating natural disasters. As we have seen, humans survived multiple bottlenecks, resulting in morphological changes associated with founder effects. There is no doubt at all that we are one species and one race. The differences in communities across vast regions are in fact minuscule, and utterly trivial.

It is my strong suspicion that the colonisation of Europe by *Homo sapiens sapiens* only became possible after modern humans gained evolutionary advantages from their interbreeding with Neanderthals, a human type that had acquired adaptations allowing them to tolerate intensely cold climates. There is no sign of our lineage in Europe until after the dates calculated for interbreeding with Neanderthals. Today all modern humans in Europe carry significant levels of DNA from these people. A great debt is owed to these tough guys and gals; they allowed many more of us to survive the depths of the last glacial maximum.

There has been no discussion of the American continental region in this book. That does not mean that it has no part in the story of ancient migrations, or that the current mainstream hypothesis is correct for that landmass. It is my understanding that Australoid groups migrated into the Americas at a very early epoch. The evidence for this claim will be assessed in the sequel to this book. It is a history of tremendous adversity, arduous journeys and a long lost civilisation now swallowed by jungles.

It is my opinion that, long after the two waves of Australasian migrants entered Africa, the third great exodus began. As the climate improved, around 60 Kya, people slowly emerged from their refuges in South Africa and Australasia. They followed green belts and migrating animals, slowly colonising the planet in earnest. It makes the most sense to imagine that these two groups encountered each other, perhaps in the Levant, maybe on the Arabian Peninsula. Wherever the meeting happened, it is only rational to assume that they were glad to find other people who had survived the grim times. Perhaps they celebrated their meeting, but if we have learned one thing about humans, we can say for sure that some of them ended up in bed together. These two migration events, one coming out of Africa and the other coming from Australasia, allowed gene flow in all directions. We represent the ultimate fusion into the global human race of today.

Human evolution can now be understood as firstly an African story, later becoming an Australasian story, but eventually, the interweaving of two threads led to a single global human story. *Homo sapiens* may well have been on their evolutionary line for more than 700 Kya, but most likely they were not recognisable as fully anatomically modern humans until closer to 400 Kya. People today stem from a single population of Australasian Aboriginals, but they are also representatives of the various hominin populations encountered along the way. We are, in the end, an ancient and glorious race of hybrids.

Sources & References

1 Foreword

griffith.edu.au/environment-planning-architecture/
environmental-futures-research-institute/research/human-evolution/
about-us

2 Chapter 1

Roberts, A, Origins of Us: Human Anatomy and Evolution,
University of Birmingham, YouTube

3 australianmuseum.net.au/homo-habilis

4 de Heinzelin et al., 1998, 'Environment and Behavior of
2.5-Million-Year-Old Bouri Hominids', ScienceVol. 284, Issue
5414, pp. 625-629

5 news.nationalgeographic.com/news/2015/03/150304-homo-
habilis-evolution-fossil-jaw-ethiopia-olduvai-gorge/

6 Harmand et al., 2015, '3.3-million-year-old stone tools from
Lomekwi 3, West Turkana, Kenya', Nature 521, 310–315

7 australianmuseum.net.au/homo-erectus

8 Stein, P & Rowe, B, 1997, Early Hominids in Asia, Physical Anthropology the Core, Chapter 11, pages 260-268

9 Wrangham, R, & Carmody, R, 2010, 'Human adaptation to the control of fire', Evolutionary Anthropology 19(5): 187–199

10 Berna et al., 'Microstratigraphic evidence of in situ fire in the Acheulean strata of Wonderwerk Cave, Northern Cape province, South Africa', 2012, vol. 109 no. 20, 7593–7594

11 Stein, P & Rowe, B, 1997, Early Hominids in Asia, Physical Anthropology the Core, Chapter 11, pages 260-268

12 Huffman et al., 2005, 'Historical Evidence of the 1936 Mojokerto Skull Discovery, East Java', Journal of Human Evolution, Volume 48, Issue 4, Pages 321-363

13 Morwood, M. J., O'Sullivan, P., Susanto, E. E. & Aziz, F. (2003). Revised age for Mojokerto 1, an early Homo erectus cranium from East Java, Indonesia. Australian Archaeology, 57 1-4.

14 Lordkipanidze, et al., 2013, 'A Complete Skull from Dmanisi, Georgia, and the Evolutionary Biology of Early Homo', Science, Vol. 342, Issue 6156, pp. 326-331

Chapter 2

15 Toro-Moyano et al., 2013, The oldest human fossil in Europe, from Orce (Spain), Journal of Human Evolution Volume 65, Issue 1, Pages 1-9

16 Toro-Moyano et al., 2011, 'The archaic stone tool industry from Barranco León and Fuente Nueva 3, (Orce, Spain): Evidence of the earliest hominin presence in southern Europe', Quaternary International, Volume 243, Issue 1, 19 October 2011, Pages 80-91

17 Huff et al., 2009, 'Mobile elements reveal small population size in the ancient ancestors of Homo sapiens', vol. 107 no. 5, 2147–2152

18 australianmuseum.net.au/homo-antecessor

19 Parés et al., 2013, 'Reassessing the age of Atapuerca-TD6 (Spain): new paleomagnetic results', Journal of Archaeological Science, Volume 40, Issue 12, Pages 4586-4595

20 Carbonell et al, 2008, 'The First Hominin of Europe', Nature 452, 465-469

21 australianmuseum.net.au/homo-heidelbergensis

22 Meyer et al., 2016, 'Nuclear DNA sequences from the Middle Pleistocene Sima de los Huesos hominins', Nature, 531, 504–507

23 Gómez-Robles et al., 2013, 'No known hominin species matches the expected dental morphology of the last common ancestor of Neanderthals and modern humans', PNAS, vol. 110 no. 45, 18196–18201

24 natureworldnews.com/articles/4549/20131022/study-known-hominin-ancestor-neanderthals-modern-humans.htm

25 nature.com/news/oldest-ancient-human-dna-details-dawn-of-neanderthals-1.19557

26 nutcrackerman.com/2016/11/09/a-moment-of-silence-for-the-death-of-homo-heidelbergensis/

Chapter 3

27 Ahmed, M & Liang, P, 2013, 'Study of Modern Human Evolution via Comparative Analysis with the Neanderthal Genome', Genomics Information, 11(4): 230–238

28 Dalén et al., 2012, 'Partial Genetic Turnover in Neandertals: Continuity in the East and Population Replacement in the West', Molecular Biology Evolution 29 (8): 1893-1897

29 Meyer et al., 2012, 'A High-Coverage Genome Sequence from an Archaic Denisovan Individual', Science Vol. 338, Issue 6104, pp. 222-226

30 Reich et al, 2010, 'Genetic history of an archaic hominin group from Denisova Cave in Siberia', Nature 468, 1053–1060

31 scientificamerican.com/article/denisovan-genome/

32 johnhawks.net/weblog/reviews/neandertals/neandertal_dna/
sima-de-los-huesos-dna-meyer-2013.html

33 Wall et al., 2013, 'Higher Levels of Neanderthal Ancestry in
East Asians Than in Europeans', Genetics, vol. 194 no. 1 199-209

34 Meyer et al., 2012, 'A High-Coverage Genome Sequence from
an Archaic Denisovan Individual', Science Vol 338

35 anthropogenesis.kinshipstudies.org/blog/2012/03/17/
american-indians-neanderthals-and-denisovans-pca-views/

36 archaeology.org/news/2455-140821-europe-neanderthal-
extinction-redated

37 Racimo et al., 'Evidence for archaic adaptive introgression in
humans', Nature Reviews Genetics 16, 359–371

38 sciencemag.org/news/2016/02/humans-mated-neandertals-
much-earlier-and-more-frequently-thought

39 bbc.com/earth/story/20150122-is-this-a-new-species-of-human

40 Wu et al., 2014, 'Temporal labyrinths of eastern Eurasian
Pleistocene humans', PNAS vol. 111 no. 29, 10509–10513

41 Bae et al., 2014, 'Modern human teeth from Late Pleistocene Luna Cave (Guangxi, China)', Quaternary International Volume 354, Pages 169-183

42 Liu et al., 2010, 'Human remains from Zhirendong, South China, and modern human emergence in East Asia', vol. 107 no. 45, 19201–19206

43 nature.com/nature/journal/v526/n7575/full/nature15696.html

44 Hershkovitz et al., 2015, 'Levantine cranium from Manot Cave (Israel) foreshadows the first European modern humans', Nature 520, 216–219

45 nature.com/news/how-china-is-rewriting-the-book-on-human-origins-1.20231

46 Kimura et al., 2009, 'A Common Variation in EDAR Is a Genetic Determinant of Shovel-Shaped Incisors', Am J Hum Genet. 85(4): 528–535.

47 anthro.palomar.edu/homo2/mod_homo_4.htm

48 scmp.com/news/china/society/article/1987039/178000-years-chinese-history-thats-really-something-chew

49 Sankararaman et al., 2016, 'The Combined Landscape of Denisovan and Neanderthal Ancestry in Present-Day Humans', Current Biology, Volume 26, Issue 9, p1241–1247

Chapter 4

50 self.gutenberg.org/articles/Eug%C3%A8ne_Dubois

51 archaeologyhub.info/indonesian-pyramid-is-20000-years-old-claims-geologist-discovery-may-rewrite-history/

52 britannica.com/topic/Solo-man

53 peterbrown-palaeoanthropology.net/Stone Tools on Flores 1.htm

54 Brown et al., 2004, 'A new small-bodied hominin from the Late Pleistocene of Flores, Indonesia', Nature 431, 1055-1061

55 Brumm et al., 2016, 'Age and context of the oldest known hominin fossils from Flores', Nature 534, 249–253

56 livescience.com/55014-miniature-hobbit-ancestors-discovered.html

57 australianmuseum.net.au/homo-floresiensis

58 Stein, P & Rowe, B, 1997, Early Hominids in Asia, Physical Anthropology the Core, Chapter 11, pages 260-268

59 nature.com/news/did-humans-drive-hobbit-species-to-extinction-1.19651

60 phys.org/news/2015-12-bone-red-deer-cave-people.html

61 smithsonianmag.com/science-nature/were-the-hobbits-ancestors-sailors-1231030/

62 Madsen, J, 2012, 'Who was Homo erectus', Natural History Museum of Denmark, Science Illustrated, p. 23

63 Strasser et al., 2010, 'Stone Age Seafaring in the Mediterranean: Evidence from the Plakias Region for Lower Palaeolithic and Mesolithic Habitation of Crete'Hesperia: The Journal of the American School of Classical Studies at Athens, Vol. 79, No. 2, pp. 145-190

64 Van den Bergh et al., 2016, 'Earliest hominin occupation of Sulawesi, Indonesia', Nature 529, 208–211

65 labs.iro.umontreal.ca/~vaucher/History/Prehistoric_Craft/

66 Luke-Killam, A, 2001, 'Language Capabilities of Homo erectus & Homo neanderthalensis', LIGN 272

67 Benítez-Burraco, A, & Barceló-Coblijn, L, 2013, 'Paleogenomics, hominin interbreeding and language evolution', Journal of Anthropological Sciences, Vol. 91, pp. 239-244

68 Wyn et al. 'The Origins of Speech', Cambridge Archaeological Journal, Volume 8, Issue 1, pp. 69-94

69 Knight et al., 2003, 'African Y chromosome and mtDNA divergence provides insight into the history of click languages', Current Biology, 13(8):705

70 Whitley, Life and Letters of Charles Darwin, (Vol 1, pp. 226–27)

71 Joordens et al., 2015, 'Homo erectus at Trinil on Java used shells for tool production and engraving', Nature 518, 228–231

72 australianarchaeologicalassociation.com.au/media-releases/ice-age-living-in-the-pilbara/

Chapter 5

73 Henshilwood et al., 2009, 'Engraved ochres from the Middle Stone Age levels at Blombos Cave, South Africa', Journal Human Evolution; 57(1):27-47

74 joannenova.com.au/2010/02/the-big-picture-65-million-years-of-temperature-swings/

75 ces.fau.edu/nasa/module-3/temperature-changes/exploration-1.php

76 Bednarik, R, 200, 'Pleistocene Timor: some corrections', Australian Archaeology
No. 51, pp. 16-20

77 O'Connor, S, Ono, R & Clarkson, C, 2011, 'Pelagic Fishing at 42,000 Years Before the Present and the Maritime Skills of Modern Humans', Science Vol. 334, Issue 6059, pp. 1117-1121

78 theaustralian.com.au/higher-education/asian-neanderthals-may-have-occupied-australia/news-story/4c135f17785ccb8b2e0 cb6241687c093

79 Sankararaman, S, 2016, 'The Combined Landscape of Denisovan and Neanderthal Ancestry in Present-Day Humans', Current Biology Volume 26, Issue 9, p1241–1247

80 abc.net.au/news/science/2016-09-22/world-first-study-reveals-rich-history-of-aboriginal-australians/7858376

81 'Aboriginal Genome Reveals New Insights into Early Humans', Australian Science, November 2011

82 sciencemag.org/news/2016/02/humans-mated-neandertals-much-earlier-and-more-frequently-thought

83 ibid

84 Malaspinas, A, 2016, 'A genomic history of Aboriginal Australia', Nature 538, 207–214

85 Bohlender, R, 2016, 'A complex history of archaic admixture in modern humans', ASHG 2016 Meeting

86 pangeanic.com/knowledge_center/country-with-the-highest-level-of-language-diversity-papua-new-guinea/

87 Malaspinas, A, 2016, 'A genomic history of Aboriginal Australia', Nature 538, 207–214

88 ibid

Chapter 6

89 smh.com.au/news/national/when-i-was-fauna-citizens-rallying-call/2007/05/22/1179601412706.html

90 Wilson, A & Cann, R, 1992, 'The Recent African Genesis of Humans', Scientific American 266(4):68-73.

91 Gribbin, J & Cherfas, J, 1982, 'The Monkey Puzzle: Reshaping the Evolutionary Tree', Pages 247-251

92 Gribbin, J & Cherfas, J, 1982, 'The Monkey Puzzle: Reshaping the Evolutionary Tree', Pages 253

93 Gribbin, J & Cherfas, J, 1982, 'The Monkey Puzzle: Reshaping the Evolutionary Tree', Pages 254

94 Cann, R, Stoneking, M & Wilson, A, 1987, 'Mitochondrial DNA and Human Evolution', Nature 325, 31 - 36

95 Chakraborty, R, 1990, 'Mitochondrial DNA Polymorphism Reveals Hidden Heterogeneity within Some Asian Populations', American Journal of Human Genetics 47:87-94

96 Excoffier, L & and Langaney, A, 1989, 'Origin and Differentiation of Human Mitochondrial DNA', Am. J. Hum. Genet. 44:73-85

97 von Platen, J, 2007, A history and interpretation of fire frequency in dry eucalypt forests and woodlands of Eastern Tasmania', University of Tasmania

98 Hiscock, P & Kershaw, P, 1992, 'Palaeoenvironments and prehistory of Australia's tropical Top End', ANU Research Publications

99 moyjil.com.au/how-old

100 'Not Out of Africa - Alan Thorne's challenging ideas about human evolution', Discover Magazine, August 2002 Issue

101 australiangeographic.com.au/blogs/on-this-day/2013/02/on-this-day-mungo-man-fossil-found/

102 Cohen et al., 2015, 'Hydrological transformation coincided with megafaunal extinction in central Australia', Geological Society of America

103 http://cogweb.ucla.edu/ep/Mungo_Man.html

104 Adcock et al., 2001, 'Mitochondrial DNA sequences in ancient Australians: Implications for modern human origins', PNAS vol. 98 no. 2, 537–542

105 Heupink et al., 2016, 'Ancient mtDNA sequences from the First Australians revisited', PNAS, vol. 113 no. 25, 6892–6897

106 abc.net.au/news/2016-06-07/dna-confirms-aboriginal-people-as-the-first-australians/7481360

107 peterbrown-palaeoanthropology.net/KowS.html

108 Grün R et al., 2011, 'The chronology of the WLH 50 human remains, Willandra Lakes World Heritage Area, Australia', Journal of Human Evolution, 60(5):597-604

Chapter 7

109 australianmuseum.net.au/the-first-modern-humans-in-southeast-asia

110 Rasmussen et al., 2011, 'An Aboriginal Australian genome reveals separate human dispersals into Asia', Science, 7;334(6052):94-8

111 'DNA confirms Aboriginal culture one of Earth's oldest', Australian Geographic, September 2011

112 sciencemag.org/news/2016/09/almost-all-living-people-outside-africa-trace-back-single-migration-more-50000-years

113 bradshawfoundation.com/bradshaws/introduction.php

114 isogg.org/wiki/Haplogroup

115 en.wikipedia.org/wiki/Macro-haplogroup_L_(mtDNA)

116 dienekes.blogspot.com.au/2011/11/age-of-mtdna-haplogroup-l3-about-70.html

117 Soares et al., 2012, 'The Expansion of mtDNA Haplogroup L3 within and out of Africa', Mol Biol Evol. 29(3):915-27

118 sapiens.org/evolution/early-human-migration/

119 voices.nationalgeographic.org/2015/12/16/new-genographic-research-from-australia-discovers-unique-branches-of-the-human-family-tree/

120 revolvy.com/main/index.php?s=Haplogroup%20C-M130%20(Y-DNA)&item_type=topic

121 Kayser et al., 2001, 'Independent Histories of Human Y Chromosomes from Melanesia and Australia', American Journal Human Genetics, 68(1): 173–190

122 le.ac.uk/departments/emfpu/to-be-deleted/explained/mitochondrial

123 Zhong et al., 2010, 'Global distribution of Y-chromosome haplogroup C reveals the prehistoric migration routes of African exodus and early settlement in East Asia', Journal of Human Genetics, 55, 428–435

124 Fu et al., 2014, 'Genome sequence of a 45,000-year-old modern human from western Siberia', Nature 514, 445–449

125 Green et al., 'A Draft Sequence of the Neandertal Genome', Science Vol. 328, Issue 5979, pp. 710-722

126 Bazgir et al., 'Understanding the emergence of modern humans and the disappearance of Neanderthals: Insights from Kaldar Cave (Khorramabad Valley, Western Iran)', Scientific Reports 7, Article number: 43460

127 Roberts, R, 2012, 'Armageddon and its aftermath: dating the Toba super-eruption', Uni. of Wollongong Research Online

128 Ambrose, S, 1998, 'Late Pleistocene human population bottlenecks, volcanic winter, and differentiation of modern humans', Journal of Human Evolution, 34, 623-651

129 sites.google.com/site/tobavolcano/

130 andamans.org/toba-aftermath-climate-and-environment/

131 Petraglia et al., 2007, 'Middle Paleolithic Assemblages from the Indian Subcontinent Before and After the Toba Super-Eruption', Science Vol. 317, Issue 5834, pp. 114-116

132 'IMPACT Event: The First Genomic Study of Indigenous Australia', Griffith University, Griffith Sciences Youtube

Chapter 8

133 Malaspinas et al., 2016, 'A genomic history of Aboriginal Australia', Nature 18299

134 cambridgeblog.org/2013/11/peter-forster-on-archaeogenetic-breakthroughs-dna-prehistory/

135 Lane, C, Chorn, B & Johnson, T, 2013, 'Ash from the Toba supereruption in Lake Malawi shows no volcanic winter in East Africa at 75 ka', PNAS 110(20): 8025–8029

136 Ambrose, S, 1998, 'Late Pleistocene human population bottlenecks, volcanic winter, and differentiation of modern humans', Journal of Human Evolution, 34, 623-651

137 robertlindsay.wordpress.com/2009/05/05/the-birth-of-the-caucasian-race/

138 io9.gizmodo.com/5879991/the-scientists-behind-mitochondrial-eve-tell-us-about-the-lucky-mother-who-changed-human-evolution-forever

139 Kim et al, 2014, 'Khoisan hunter-gatherers have been the largest population throughout most of modern-human demographic history', Nature Communications 5, Article number: 5692

140 Mallick et al., 2016, 'The Simons Genome Diversity Project: 300 genomes from 142 diverse populations', Nature 538, 201–206

141 Schlebusch et al., 2012, 'Genomic Variation in Seven Khoe-San Groups Reveals Adaptation and Complex African History', Science 1227721

142 'When the Sea Saved Humanity', Scientific American December 2012

Chapter 9

143 Hammer et al., 2011, 'Genetic evidence for archaic admixture in Africa', PNAS vol. 108 no. 37, 15123–15128

144 Pagani et al., 2016, 'Genomic analyses inform on migration events during the peopling of Eurasia', Nature 538(7624):238-242

145 'First Madagascar settlers may have been Indonesian', New Scientist magazine March 2012

146 anthropology.net/2008/07/08/the-age-of-omo-i-and-omo-ii-from-the-kibish-formation-omo-valley-ethiopia/

147 Shi et al., 2008,'Y chromosome evidence of earliest modern human settlement in East Asia and multiple origins of Tibetan and Japanese populations', BMC Biology

148 Mondal et al., 2016, 'Genomic analysis of Andamanese provides insights into ancient human migration into Asia and adaptation', Nature

149 Aghakhanian et al., 2015, 'Unravelling the Genetic History of Negritos and Indigenous Populations of Southeast Asia', Genome Biology Evolution 7(5): 1206–1215.

150 Department of Environmental Affairs of the Republic of South Africa, 2015, 'The Emergence of Modern Humans: The Pleistocene occupation sites of South Africa', UNESCO

151 http://anthro.palomar.edu/vary/vary_3.htm

CPSIA information can be obtained
at www.ICGtesting.com
Printed in the USA
FSOW03n1935190218
44800FS